Brooklands Books

☆ A BROOKLANDS ☆
'ROAD TEST' LIMITED EDITION

HUDSON 1946-1957

Compiled by
R.M.Clarke

ISBN 1 85520 4355

BROOKLANDS BOOKS LTD.
P.O. BOX 146, COBHAM,
SURREY, KT11 1LG. UK

A-HUDX1

Printed in Hong Kong

CONTENTS

ACKNOWLEDGEMENTS

For more than 35 years, Brooklands Books have been publishing compilations of road tests and other articles from the English speaking world's leading motoring magazines. We have already published more than 600 titles, and in these we have made available to motoring enthusiasts some 20,000 stories which would otherwise have become hard to find. For the most part, our books focus on a single model, and as such they have become an invaluable source of information. As Bill Boddy of *Motor Sport* was kind enough to write when reviewing one of our Gold Portfolio volumes, the Brooklands catalogue "must now constitute the most complete historical source of reference available, at least of the more recent makes and models."

Even so, we are constantly being asked to publish new titles on cars which have a narrower appeal than those we have already covered in our main series. The economics of book production make it impossible to cover these subjects in our main series, but Limited Edition volumes like this one give us a way to tackle these less popular but no less worthy subjects. This additional range of books is matched by a Limited Edition - Extra series, which contains volumes with further material to supplement existing titles in our Road Test and Gold Portfolio ranges.

Both the Limited Edition and Limited Edition - Extra series maintain the same high standards of presentation and reproduction set by our established ranges. However, each volume is printed in smaller quantities - which is perhaps the best reason we can think of why you should buy this book now. We would also like to remind readers that we are always open to suggestions for new titles; perhaps your club or interest group would like us to consider a book on your particular subject?

Finally, we are more than pleased to acknowledge that Brooklands Books rely on the help and co-operation of those who publish the magazines where the articles in our books originally appeared. For this present volume, we gratefully acknowledge the continued support of the publishers of *Auto Age, Auto Sport, Auto Sport Review, Autocar, Automobile Topics, Automobile Year Book, Car Life, Cars, Classic American, Complete Road Test, Mechanix Illustrated, Modern Motor, Motor, Motor Age, Motor Life, Motor Trend, Motor World, Special Interest Autos, Road & Track, Road Test, Speed Age* and *Wheels* for allowing us to include their valuable and informative copyright stories.

We are also indebted to Bennet Miller of the Hudson-Terraplane-Essex Club (see Notice on Page 54) firstly for suggesting this book and secondly, for photographing members' cars which appear on our front and back covers.

R.M. Clarke.

Cover Photographs

Front - Bob Van Zant's 1950 Hudson Commodore

Back Top Left and Bottom Right - Bennet Miller's 1952 Hudson Hornet 7B

Back Top Right and Bottom Left - Lou Mandrich's 1948/1949 Commodore S

New front end design is the most marked change in the 1946 Hudson models. Bumpers on the new model are heavy, wide and long.

Exterior and interior styling among several new changes to be found on the 1946 Hudson models

New Appearance Features 1946 Hudson

WITH new 1946 Hudson automobiles already coming off the production line, and the new models to be distributed to the expanded distributor-dealer organization as rapidly as possible, the Hudson Motor Car Company is prepared for a capacity output as rapidly as materials are available.

Hudson's 1946 models include an entirely new front end, new exterior styling and completely new interior, embodying comfort and convenience.

At the same time, major features which enjoyed wide popularity in the 1942 models have been retained—among them the Super-Six and Super-Eight motors; Drive-Master, which does away with clutch and gear lever operation in forward driving; and hydraulic service brakes which provide an emergency mechanical braking system operating automatically from the same foot pedal.

Two distinctly different "style" lines will be built—a Super Series and a Commodore Series, both on a 121 in. wheelbase chassis. Each of these series, however, will be available in both Sixes and Eights—corresponding body models of the same series being identical except for power plant and price.

Production will start with six-cylinder models only—Eights following as soon as reconversion of eight-cylinder motor machinery can be completed. Scheduled for first delivery are 4-door sedans, to be followed by broughams and club coupes. Later, as the most urgent demand is met, 3-passenger coupes and convertibles will be added to the line.

In addition to passenger models, Hudson will start immediate production of cab pick-ups, as part of a program to meet urgent requirements for commercial vehicles of ¾-ton size.

Chief appearance change in the 1946 Hudsons is an entirely new front end design. In the center, at headlamp level, is a newly designed adaptation of the Hudson Triangle emblem, mounted on a heavy cross bar which tops the grille, and indirectly lighted from the rear. The grille itself is much heavier than in previous designs, and is cast instead of stamped. Bumpers are heavy, wide and long, with bumper guards of tubular construction. On Commodore models, bumpers are longer and extend around the side, affording greater strength and additional fender protection. Extra guards are set near the ends of the front bumper on these models.

New belt mouldings of bright metal run the length of the car and curve gracefully down at the rear. This feature is particularly well adapted to two-tone painting, which has been reversed in 1946 models, with the dark color above and the light color below.

Commodore models for 1946 carry an auxiliary belt moulding which extends the entire length of the body, with plastic Hudson Triangle emblems at the front. All models carry series identification on the hood, below the moulding.

Wide, full-length running boards are continued for 1946, but they are concealed by the flare at the bottom of the doors.

Interiors have been completely restyled in all 1946 Hudsons, and upholstery fabrics are of neutral tone—a fine shadow weave in the Super series and Bedford cord in Commodores. Airfoam seat cushions are standard in Commodore models and an extra-cost option in the Super series.

Three Thousand Miles fo

THREE of us, whose work had taken us to New York in the autumn, spent our last fortnight in search of America—in a borrowed car, a six-cylinder Hudson coupé.

We set off, north-east from Manhattan, into New York State, thence into Connecticut and on to Boston for Thanksgiving. So far our progress had been leisurely, but on November 28 the real trip began, as we turned west at the start of the 1,097 miles we were to drive in the next three days from Boston to Chicago.

It was overcast as we left Boston and soon snow began to fall. It fell steadily until we neared the beautiful Mohawk Trail in western Massachusetts, when the sun finally broke through. Green Mountain Forest behind, we entered New York State and followed the old Taconic Trail to Troy, up and down steep, mountainous roads at the northern end of the Berkshire hills. Surfaces were slippery and snow-covered and we had to go carefully, our progress punctuated with signs announcing hairpin bends. Soon it was snowing again, and as we drove along Cherry Valley Turnpike, west of Schenectady, the snow grew deeper and deeper; big drifts, and banks of earlier snow pushed to the roadside by snowploughs, glimmered out of the darkness as we passed. A faint clicking from each passing car told us that chains were in general use, and after several fruitless attempts we bought some for ourselves; driving was becoming more and more difficult on the frozen surface and we barely made the top of one hill.

It was 11 o'clock at night when we finally drove in under a red neon sign announcing: COTTAGES; and, after the usual registration formalities, took the car right up to the two-roomed, heated, wooden cabin in which we spent the night. These clusters of roadside cabins—often called "Motels" or "Autocourts" are everywhere, and along with the roadside restaurants and cafés, the countless "gas"-stations, and the open-air drive-in cinemas are characteristic of this country of motorists and immense distances.

The cabins vary considerably—this one was rather shabby and depressing, both inside and outside, and the heating system was erratic; however, each room had a comfortable double-bed and a chair, and a cubicle with washbasin, towels and soap and plenty of hot water. Later we stayed at a motel in Virginia which was far more luxurious; it was brand-new, the outside walls spotlessly white with touches of scarlet at the windows. An archway led into a patio, round three sides of which ran small semi-detached rooms, each parquet-floored and equipped with a very comfortable bed, an easy chair, a small writing table and a dressing-table—all light oak with green furnishings; a white-tiled cubicle with washbasin and shower and venetian blinds at the windows. The charge was three dollars—about fifteen shillings per person; at the other motel, two dollars; there is no extra charge for the car. One can drive up to a motel late at night, arrange to be called, and make as early a start as one wishes.

Grim Warning

The following morning the route lead over mile after mile of flat, snow-covered country under grey, snow-laden clouds. By the time we reached a huge board instructing us to Turn Right at Next Signal—Route 324: Niagara Falls and Canada, however, the sun was shining brilliantly, though the wind was icy, but as we left Niagara and turned towards Buffalo the unnerving sight of a smashed car and two bodies laid out under rugs was an added reminder that the roads were still dangerous and cautious driving very necessary.

We had our chains fixed while we ate at a diner. These places—usually known as "Al's" or "Jo's" or "Jim's" Diners—are used mainly by the long-distance truck drivers: those huge trucks, snarling by with a

Crossing the Cumberland River, in Kentucky, on a primitive free ferry.

(Right) A blanket of heavy snow covered the landscape near Niagara.

(Extreme Right) The Hudson outside one of the "motels" which are cheap and generally comfortable.

£15 By MARY EDMOND

Through Twelve States of the U.S.A. from New York to Virginia, in a Borrowed Second-hand Car

frightening rush of sound, their galaxy of red lights winking diabolically at night, are inextricably associated with our memories of motoring in America. Diners would, I suppose, correspond roughly to our "good pull-ups for carmen." They are warm, cheerful places at which one can usually eat a good meal.

On again through the snow—until the car died on us; eventually a man was found willing to help some stranded motorists at ten o'clock of a Saturday evening; he pushed us back to his garage, which was surrounded by grotesque shapes shrouded in snow—rounded, Thurber-like objects scarcely recognizable as cars. Luckily the trouble was quickly righted, as we had arranged to be in Chicago the following night and there was still a very long way to go. Fortunately, next morning, the car started at once though we had had to leave it out in the snow all night. Out of New York State and into Pennsylvania; breakfast in Erie; out of Pennsylvania and into Ohio; Lake Erie all the time stretched, a cold, black, steely expanse, a few miles to our right. Every house was fringed with huge icicles and on every roof was a slouch-hat of snow, leaning crazily over one side. After a time the snow lessened, the sun came out and we were able to take off the chains and drive at a good speed. During the early evening we entered Indiana and the area of Central Standard Time—watches one hour back; this was a welcome lengthening of the day on which we had to drive 513 miles.

Coming Back

After a three-day stay in Central Illinois we began our return journey, south-east through the flat, rich farmland of the middle west. On one very straight road we did 100 miles in 108 minutes—this included enforced deceleration through two or three villages—the Hudson doing 80 m.p.h. as though it were 60 or less. Then through southern Indiana, across into Kentucky, and next day a drive through the lovely Cumberland State Park and across the Cumberland River by free ferry to the falls. It was now several days since we had seen snow, and at the falls it was so warm and sunny that we sat for some time without coats.

JOURNEY'S END—Fifth Avenue, New York, from Central Park

Through the Cumberland Gap, at the western tip of Virginia, we came upon a most spectacular mountain road, where fold upon fold of mighty peaks faded into an indescribable sunset. The Hudson took the steep corkscrews like a bird.

Next day we reached the Skyline Drive through Shenandoah National Park, at the beginning of which is a sign: NEXT GAS STATION FORTY MILES ON. This magnificent road, constructed at an average height of about 2,500 ft., has frequent "overlooks," where the motorist may draw off the road to stop and admire the crowding mountains, the road making its sinuous way around and between them, and the valleys beyond.

And so, back through Alexandria and Washington, to New York and the end of the trip.

Road maps, of different degrees of clarity and informativeness, are supplied free of charge by state highway authorities and all the leading oil companies; maps covering the state through which one is driving and adjoining states are supplied on request by petrol stations.

And the actual cost? Expenditure on the car on the outward trip was $40.58, or just about £10. This total was made up thus: Anti-freeze on the first day; two 10-cent tolls on a parkway; oil on the third day; chains ($8); fixing a flat tyre; a toll on Grand Island bridge at Niagara; fixing the chains; the $4 charged by the man who came for us, pushed us several miles back to his garage and righted the car; garaging the car for one night in Chicago ($2), and petrol. Expenditure on the return trip, for the 1,371 miles from central Illinois to New York, was $21.55, or just over £5. This was accounted for by tolls, tunnel and ferry-boat charges, a tyre change, and petrol.

The Motor Continental Road Test No. 1C/49

Make: Hudson. **Type:** Commodore Custom 8 Club Coupé.

Makers: The Hudson Motor Car Company, Detroit, 14, Michigan, U.S.A.

Dimensions and Seating

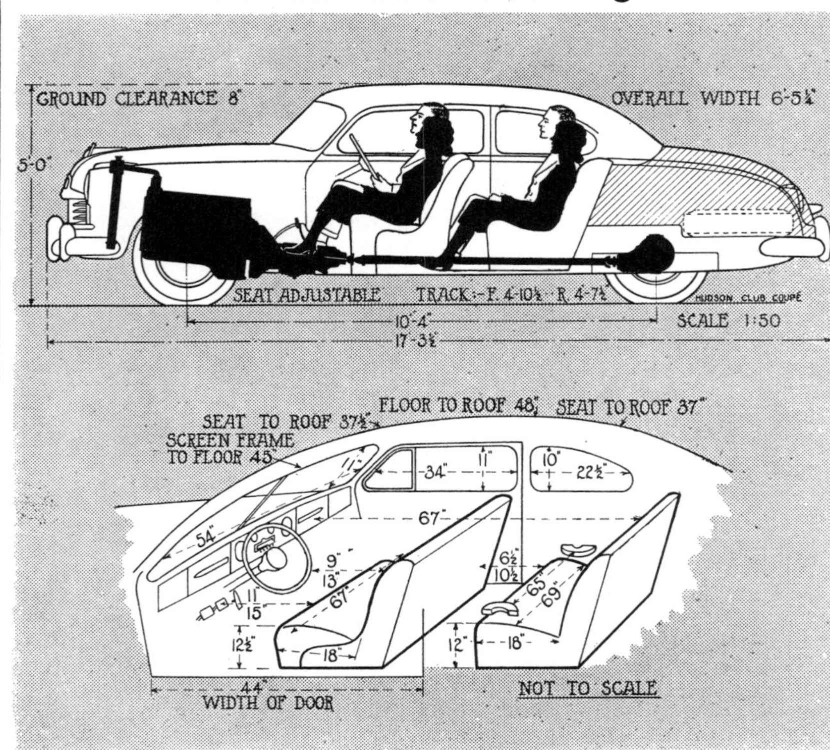

Test Conditions

Cool, slight breeze, dry concrete surface, Belgian pump uel.

Test Data

ACCELERATION TIMES on Two Upper Ratios

	Top Overdrive	Top Direct	2nd Overdrive	2nd Direct
10-30 m.p.h.	— secs.	9.1 secs.	— secs.	5.2 secs.
20-40 m.p.h.	— secs.	8.9 secs.	— secs.	5.5 secs.
30-50 m.p.h.	15.4 secs.	8.8 secs.	7.1 secs.	7.0 secs.
40-60 m.p.h.	15.8 secs.	9.3 secs.	8.4 secs.	— secs.
50-70 m.p.h.	20.4 secs.	11.9 secs.	11.9 secs.	— secs.
60-80 m.p.h.	28.2 secs.	20.7 secs.	— secs.	— secs.

ACCELERATION TIMES Through Gears

0-30 m.p.h.	5.6 secs.
0-40 m.p.h.	9.0 secs.
0-50 m.p.h.	13.6 secs.
0-60 m.p.h.	17.5 secs.
0-70 m.p.h.	24.7 secs.
0-80 m.p.h.	36.6 secs.
Standing quarter-mile	21.1 secs.

MAXIMUM SPEEDS

Flying Quarter-mile (overdrive top gear)
Mean of four opposite runs .. 90.3 m.p.h.
Best time equals .. .. .. 91.8 m.p.h.

Speed in Gears

Max. speed in top gear (direct) 84 m.p.h.
Max. speed in 2nd gear (overdrive) 77 m.p.h.
Max. speed in 2nd gear (direct) 58 m.p.h.
Max. speed in 1st gear (overdrive) 52 m.p.h.
Max. speed in 1st gear (direct) 35 m.p.h.

BRAKES AT 30 m.p.h.

0.33 g (= 91 ft. stopping distance) with 25 lb. pedal pressure.
0.68 g (= 44½ ft. stopping distance) with 50 lb. pedal pressure.
0.91 g (= 33 ft. stopping distance) with 70 lb. pedal pressure.

FUEL CONSUMPTION

Overall consumption approx. 16 m.p.g.
24.5 m.p.g. at constant 30 m.p.h.
22.5 m.p.g. at constant 40 m.p.h.
21.0 m.p.g. at constant 50 m.p.h.
19.0 m.p.g. at constant 60 m.p.h.
17.5 m.p.g. at constant 70 m.p.h.
15.5 m.p.g. at constant 80 m.p.h.

STEERING

Left-hand lock 41 ft.
Right-hand lock 42 ft.
5¾ turns of steering wheel, lock to lock.

HILL CLIMBING

Max. top gear speed on 1 in 20 74 m.p.h.
Max. top gear speed on 1 in 15 68 m.p.h.
Max. top gear speed on 1 in 10 55 m.p.h.
Max. gradient climbable on overdrive top gear 1 in 13½ (Tapley 170 lb. per ton).
Max. gradient climbable on direct top gear 1 in 8¼ (Tapley 275 lb. per ton).
Max. gradient climbable on overdrive 2nd gear 1 in 7½ (Tapley 300 lb. per ton).
Max. gradient climbable on direct 2nd gear 1 in 5 (Tapley 450 lb. per ton).

In Brief

Price $2,486 f.o.b. Detroit, with extra equipment but excluding Federal Tax. Equivalent price at £1 equals $4.03 equals £616¾.

Capacity	4,160 c.c.
Road weight unladen	33½ cwt.
Front/rear wt. distribution	57/43
Laden weight as tested	37 cwt.
Fuel consumption	16 m.p.g.
Maximum speed	90.3 m.p.h.
Maximum speed on 1 in 20 gradient	74 m.p.h.
Maximum top gear gradient (direct drive)	1 in 8¼
Acceleration, 10-30 on top (direct drive)	9.1 secs.
0-50 through gears	13.6 secs.

Gearing 25.2. m.p.h. in overdrive top at 1,000 r.p.m. 84 m.p.h. at 2,500 ft. per minute piston speed in overdrive top gear.

Specification

Engine

Cylinders	Straight 8
Bore	76.2 mm.
Stroke	114.3 mm.
Cubic Capacity	4,160 c.c.
Piston Area	56.5 sq. ins.
Valves	Side
Compression Ratio	6.5
Max. power	128 b.h.p.
at	4,000 r.p.m.
B.h.p. per sq. in. piston area	2.26
Piston speed at max. b.h.p.	3,000 ft./min.
Carburetter	Carter dual downdraught
Ignition	Auto-lite coil
Sparking plugs	14 mm. Champion J9
Fuel pump	Mechanical
Oil filter	Optional by-pass

Transmission

Clutch	Oil immersed 10-in. single plate
Top gear overdrive	3.27
Top gear direct	4.55
2nd gear overdrive	5.96
2nd gear direct	8.29
1st gear overdrive	9.41
1st gear direct	13.1
Propeller shaft	2 in. tubular divided
Final drive	Hypoid bevel

Chassis

Brakes	Bendix hydraulic
Brake drum diameter	11 ins.
Friction lining area	158.7 sq. ins.
Tyres	Goodyear 7.60 x 15 (normal size 7.10 x 15 728 revs./mile)
Steering Gear	Gemmer worm and roller

Performance factors (at laden weight as tested)

Piston area, sq. ins. per ton	30.6
Brake lining area, sq. ins. per ton	86
Litres per ton-mile (direct top gear)	3,275

Maintenance

Fuel tank: 16½ gallons. **Sump:** 11½ pints, S.A.E.20 winter, S.A.E.30 summer. **Gearbox:** 1¾ pints, E.P. gear oil S.A.E.80 winter, S.A.E. 90 summer. **Overdrive:** 2¾ pints, E.P. gear oil S.A.E. 80 winter, S.A.E. 90 summer. **Rear axle:** 3 pints, S.A.E. 90 hypoid gear oil. **Radiator:** 30 pints. **Chassis lubrication:** Grease gun to 32 points every 1,000 miles. **Ignition timing:** 1-8 timing mark on flywheel. **Spark plug gap:** 0.032 in. **Contact breaker gap:** 0.017 ins. **Tappets (hot):** inlet 0.006 in., exhaust 0.008 in. **Front wheel toe-in:** 1/32 in. **Camber angle:** ½°. **Castor angle:** 0° **Kingpin inclination:** 3° 36'. **Tyre Pressures:** Front 24 lb., Rear 24 lb. **Brake fluid:** Hudson. **Shock absorber fluid:** No provision for refilling. **Battery:** 6 volt, 120 amp.-hour. **Lamp bulbs:** Headlamps—sealed beam, parking/signal and tail/stop lamps—3/21 c.p., number-plate lamp 3 c.p., front interior lamp 15 c.p., rear interior lamps 7 c.p., door-step lamps 15 c.p., miscellaneous lamps 2 c.p.

Ref. U.S./42/49.

The HUDSON COMMODORE 8 Club Coupé

A Distinctive 90 m.p.h. American Car With a Transmission Providing Two-pedal or Normal Control

On their introduction about a year ago, the latest Hudson models created something of a sensation by the enclosure of the rear wheels, not merely within the body panelling in the manner now quite common, but within the actual box-section side members of the basic frame on which the combined body-cum-chassis structure is built up. By this device, the makers were able to use a frame structure of exceptional width and thus drop the whole floor and seat level and, of course, lower the entire body without sacrifice of head-room.

Exceptionally good styling was combined with this basic change to produce an unusually sleek appearance and good aerodynamic form. The change did, however, involve some slight narrowing of the rear track in relation to the front and the problematical effect of this on handling made the latest Hudson a car which we were particularly anxious to test.

Another distinctive Hudson feature of more than ordinary interest is the optional Drive Master transmission which is offered as an extra in addition to the well-known Warner overdrive, which is also an optional fitting.

Revising First Impressions

Thanks to the courtesy of the English Hudson concern and the co-operation of the Belgian distributors, we were recently able to satisfy ourselves on all these points. The car placed at our disposal was a straight-eight model with a close-coupled Club Coupé body of the Commodore (de luxe) series.

First impressions are always interesting, although not always correct. The latter qualification was operative in the case of the Hudson. Our first acquaintance with it was in the course of a short run after dark and, when both drivers had put in a few miles at the wheel, both confessed to a slight feeling of disappointment that the performance did not seem quite so lively as the specification had suggested.

In truth, that initial impression was one of the biggest tributes to the car, since the performance was, in fact, quite up to expectation, and the disappointment was brought about by the very notable smoothness and silence of the Hudson, aided and abetted in some measure by that rare fitting, a slightly slow speedometer. The fact that the full performance of the car—revealed the following day by the dispassionate evidence of stop watches—had not been appreciated was also a tribute to general handling qualities, as the misjudgment led to no awkward moments.

Both this initial run, and later experience over varied road conditions, revealed that the Hudson handled particularly well and that the fundamental change of rear axle layout in relation to the frame has certainly produced no ill effects. On corners, the car may be graded as excellent by touring car standards. There is practically no roll at any time (anti-roll bars are fitted at both ends) and the car takes 50-60 m.p.h. corners in its stride with no trace of unsteadiness and no call for heroics on the part of the driver.

The actual manual effort required on the steering wheel is, perhaps, higher than one would expect in view of the comparative low gearing, which entails $5\frac{3}{4}$ turns of the wheel from lock to lock; in fairness, however, it must be added that the model in question was fitted with oversize tyres and also that these $5\frac{3}{4}$ turns go to the production of a large wheel deflection and correspondingly small turning circle for so large a car. Indeed, the Hudson causes surprises in the spaces in which it can be parked and the tight turns which can be negotiated without reversing.

The steering, moreover, is entirely devoid of kick-back through the wheel, even when extremely bad pavé is taken fast, and surfaces of this kind produce no tendency to wandering.

Equally satisfactory on pavé is the suspension system which gives a remarkably even ride with no trace of pitching or vertical float. Obviously, the Hudson is very well damped and is, in fact, at its best at high speeds. At low and medium speeds, it still provides very good insulation from actual bumps, but the fact that it is not entirely proof against vibrations of moderate frequencies was shown up by a facia-board rattle which sometimes occurred under these conditions—and which, incidentally, led to the discovery that some particularly attractive grained "woodwork" about the facia board was, in fact, unusually well finished metal.

As previous remarks have already indicated, the straight-eight, 4,160 c.c. side-valve engine is extremely smooth and quiet, and to these good qualities may be added an absence of any tendency to running-on and an almost complete absence of pinking, although whether the absence of these tendencies would be equally marked on British Pool petrol is a matter for speculation.

Flexible, Smooth and Fast

As one would expect from an engine of this type and size, flexibility is a noticeable feature; it is, however, interesting to study the direct top-gear acceleration figures which clearly show that no efforts have been made to concentrate on the lower part of the power curve, the best acceleration being noticeable in the middle speed range, the 30-50 m.p.h. time of 8.8 secs. comparing with a 10-30 m.p.h. figure of 9.1 secs.

Throughout the entire range the engine is effortless and, whilst it is, perhaps, invidious to name any particular cruising speed in a car of this kind, it is time to say that a genuine 70 m.p.h. is a speed which any driver, moderately anxious to get from A to B, would be likely to hold for long periods over main roads, the more so as at high speeds the general level of wind and wheel noise is low, the latter point particularly noticeable on the stone setts of Belgian roads which set up a very pronounced under-chassis noise on any car. In this connection, the excellent built-in radio set (with loud speaker

CONTROL IN COMFORT.—A two-spoke steering-wheel, sensitive horn-ring, centrally placed instruments and good detail finish characterize the forward interior. Three pedals are provided, but use of the clutch is optional.

SPACE TO SPARE.—The Hudson's coupé body offers a magnificent amount of luggage room in the long, swept tail—even though the spare wheel takes up some of the space.

dropped flush into the considerable horizontal portion of the scuttle between the top of the facia and the steeply raked and curved Vee screen) provided a useful sound meter. With many cars, a volume which suits high-speed cruising proves very unpleasantly loud when the speed is reduced for passing through a village; no such phenomena was noticeable with the Hudson.

Automatic Alternatives

This quiet running is, of course, more marked with the over-drive in operation —a remark which automatically brings us to a consideration of the transmission choices open to owners of these cars equipped with both optional extras.

There are, in effect, three basic systems at the disposal of the driver, and each may be used with or without the over-drive, making, in effect, six combinations in all. Admittedly, this sounds complicated but, in practice, one soon learns what is which and comes to choose suitably according to taste and conditions.

The first basic state is with the Drive Master knob in the central position and the device out of action. In this condition, the three-speed gearbox operates in an entirely normal manner and one uses the clutch and steering-column lever in the ordinary way.

Turning the Drive Master knob to the left gives a partially automatic transmission in the shape of clutchless gear changing. That is to say, one starts from rest in the normal way and thereafter ignores the clutch except for actually stopping and restarting, gear-changing being effected simply by moving the gear lever to the appropriate position, the clutch being withdrawn automatically by a vacuum control. This control does not give a "free wheeling" effect as a centrifugal governor cuts it out above 20 m.p.h. in top gear.

Third choice is to turn the control knob to the right, which gives a fully automatic two-speed (top and second) transmission. To start, the driver merely places the gear lever into the top-gear position and depresses the accelerator pedal, when the car moves off in second gear. Any desired speed can be obtained in that gear merely by continuing to accelerate. So soon, however, as the accelerator pedal is released after reaching the pre-determined changing speed, the transmission automatically changes itself into top.

On slowing down with the accelerator released, the second gear becomes automatically engaged when the car speed falls below the changing speed. There is no need to declutch for a stop, control being entirely by accelerator pedal.

Finally, there is the over-drive which provides a 28 per cent. reduction in engine speed compared to direct top gear and is semi-automatic in action; that is to say, no manual operation is involved, but the action of the device is at all times under the control of the driver by means of the accelerator pedal. At any speed above approximately 22 m.p.h., the accelerator pedal has only to be released momentarily for the over-drive to engage and to remain in engagement until the speed falls below that figure, when reversion takes place to direct top.

The beauty of the device, however, is that a "kick-down" change from over-drive to direct drive is available at any speed simply by depressing the accelerator pedal fully, when the change takes place automatically and without any check. The over-drive, incidentally, incorporates a free wheel which operates at speeds below about 19 m.p.h., but both the over-drive and the free wheel can be put out of action simply by pulling out the over-drive knob—an arrangement which overcomes any safety objections.

Reverting to the Drive Master, we found the clutchless change setting effective but felt that the saving in effort (in view of the excellent synchromesh) not very great. The fully automatic setting, on the other hand, did enable one to forget gear changing entirely and to indulge in two-pedal motoring. No doubt this has a distinct appeal to drivers who wish to put a minimum of thought and effort into car control but, for ourselves, we preferred the more lively results obtained by the little extra trouble of using the transmission as a normal three-speed gearbox in conjunction with the over-drive, the more so as we found that smoother results were obtained in this way. The latter comment, however, is not a fair one without adding the qualification that the particular model we tried had been standing unused for some months and that it is possible that the mechanism was not operating at its full efficiency.

Of the over-drive, however, we cannot speak too highly as it enabled all the advantages of an ultra-high top in the matter of effortless and economical running to be obtained without any of the drawbacks, since a little extra pressure on the accelerator enabled the lower direct drive to be obtained immediately for overtaking— a quality which showed up to particular advantage when returning to Brussels with the usual Sunday evening traffic.

Summary of Details

Inevitably, consideration of an elaborate transmission system such as this occupies a good deal of space and consideration of features of the Hudson not already mentioned must be brief. To summarize, we found visibility good in the main and the sight lines provided by the bonnet motif and wing mouldings useful in placing the car, but the view to extreme right and left somewhat obstructed by the sweep down of the roof to the screen pillars . . . the horn ring proved extremely handy but a little too accessible to accidental touches when manœuvring . . . the lights were up to normal American standards, which imply a good spread but rather limited range . . . the seating was comfortable and particularly adequate for three-abreast motoring . . . the general standard of interior finish earned praise and so did the comprehensive equipment which, amongst many details, included a useful interior "dome light" over the screen.

In all, we were left in no doubt that the unusual construction of this Hudson entirely justifies itself and that, with its genuine 90 m.p.h. maximum, the car must be considered as a very fine fast touring machine which in no way falls short of the performance suggested by its sleek and attractive lines.

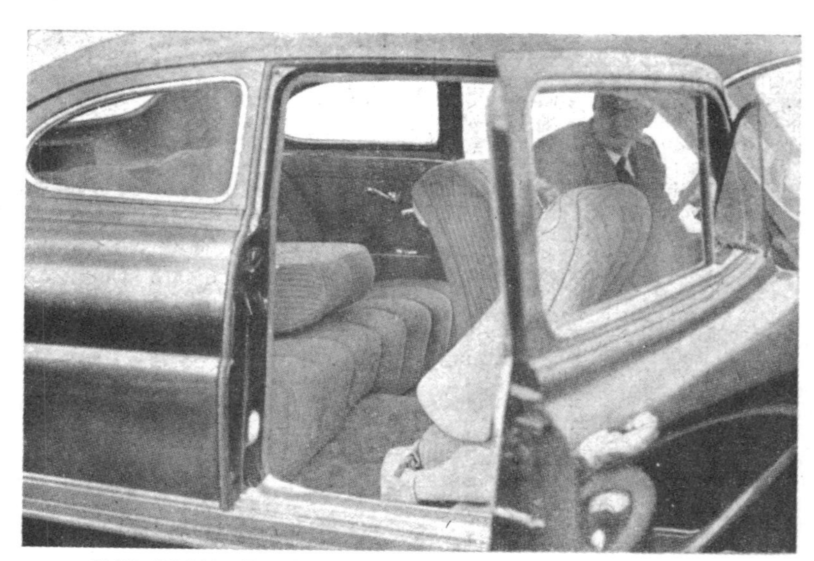

CLUB ROOM.—The Commodore Club Coupé is a genuine 6-seater

The style is modern and flush-sided but both front and rear wings are easily removable for repair. There is also a substantial plated rubbing strip.

No. 1380
HUDSON
COMMODORE
SALOON

The Autocar ROAD TESTS

DATA FOR THE DRIVER

HUDSON COMMODORE

PRICE, with four-door saloon body, $2,360, basic list price at Detroit: With additional equipment as tested, $2,733 at Detroit = £683.
RATING: 28.8 h.p., 8 cylinders, side valves, 76.2 × 114.3 mm, 4168 c.c.
BRAKE HORSE-POWER: 128 at 4,200 r.p.m. COMPRESSION RATIO: 6.5 to 1.
MAX. TORQUE: 198 lb ft at 1,600 r.p.m. 18 m.p.h. per 1,000 r.p.m. on top gear.
WEIGHT: 34 cwt 1 qr 14 lb (3,850 lb). LB per C.C.: 0.92. B.H.P. per TON: 74.47.
TYRE SIZE: 7.60 × 15in on bolt-on steel disc wheels. LIGHTING SET: 6-volt.
TANK CAPACITY: 16½ Imp. galls.; approx. fuel consumption range, 15-17 m.p.g.
TURNING CIRCLE: (L) 40ft 10in; (R) 42ft 4in. MIN. GROUND CLEARANCE: 8in.
MAIN DIMENSIONS: Wheelbase, 10ft 4in. Track, 4ft 11½in (front); 4ft 8in (rear). Overall length, 17ft 3½in; width, 6ft 5in; height, 5ft 0in.

ACCELERATION

From steady m.p.h. of

Overall gear ratios	10 to 30 sec	20 to 40 sec	30 to 50 sec	30 to 50 O.D. sec
4.55 to 1	8.0	8.1	8.9	13.2
7.50 to 1	4.7	5.3	7.0	8.8
11.80 to 1	3.6	—	—	—

From rest through gears to :—	sec
30 m.p.h.	4.8
50 m.p.h.	11.5
60 m.p.h.	18.4
70 m.p.h.	25.2
80 m.p.h.	36.5

Steering wheel movement from lock to lock: 5½ turns.

Speedometer correction by Electric Speedometer :—

Car Speedometer	Electric Speedometer	Car Speedometer	Electric Speedometer
10 =	10.5	50 =	48
20 =	20.25	60 =	57
30 =	29.75	70 =	67
40 =	39	80 =	77
		90 =	87

Speeds attainable on gears (by Electric Speedometer)

	M.p.h. (normal and max.)		M.p.h. (max.)
1st	16—33	O.D. 1st	48
2nd	40—56	O.D. 2nd	79
Top	91	O.D. = Overdrive	

WEATHER: Dry, warm; wind negligible.
Acceleration figures are the means of several runs in opposite directions.

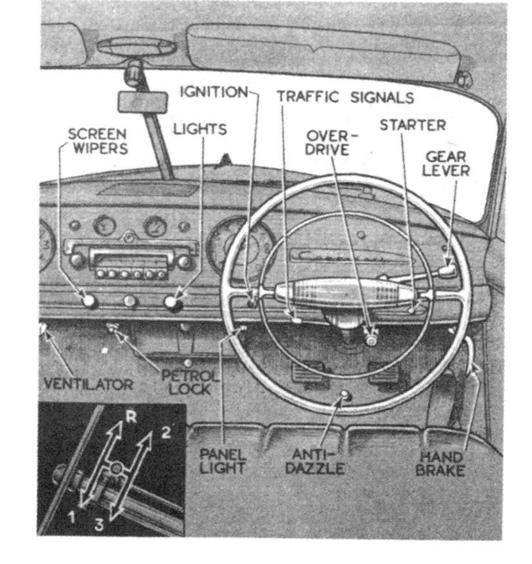

EVIDENCE is provided by the current Hudson Eight to emphasize the fact, already recognized, that the more modern designs of American cars are following an individual line, and are not necessarily just being made to look different in the interests of that overworked word, "styling." The Hudson is decidedly low built even by European standards and, in conjunction with a steel body shell-cum-frame construction, has the unique feature that the rear wheels are *inside* the outer box-section girders. To an English observer it is interesting knowledge that Mr. Reid A. Railton, British designer before the war of John Cobb's land-speed record holder, whose name was also given to the pre-war Railton car using Hudson components, is nowadays a consultant to the Hudson firm in the U.S.A.

The straight eight side-valve engine, of rather more than 4.1 litres, is slightly smaller, curiously, than the companion and otherwise identical Super Six. Independent front suspension is by coil springs and the three-speed gear box is available optionally with an overdrive transmission, as fitted to the car tested, giving, in effect, a four-speed gear box with semi-automatic changing between the third and fourth speeds. Arrangements for testing the car were made with Hudson Motors, Ltd., Great West Road, London, W.4, the British representatives, although in common with other U.S. cars the Hudson is not at present available in the ordinary way on the British market.

In first impressions of the car, apart from its striking appearance, resulting largely from the low overall height of 5ft, the most pronounced within a short distance on the road is of the light and accurate control and the stability—the steady, level way in which it rides and takes bends. Although the steering is quite low geared even by American standards, this fact is not specially apparent in ordinary driving, and is realized only when a quick swerve has to be made or when a considerable angle of movement is required at low speed in traffic. The car does not have to be held consciously on a straight course. Road wheel movements are not transmitted to the steering wheel, and the considerable steering-wheel movement that has to be made in taking a right-angle corner is greatly compensated for by the way in which the wheel spins back under the influence of strong castor action after the turn has been made.

The big eight-cylinder engine is typically self-effacing and provides an almost entirely top gear performance, inclusive of quite severe gradients and the majority of traffic

The floor of the luggage compartment is dominated by the spare wheel, but there is no question of the generous total space provided, in both width and depth. The lid is spring balanced and self-supporting.

(Right) The individual panels of the bow windscreen are curved. The radio aerial is seen in the lowered position, resting on the central pillar of the windscreen ; it is controllable from inside the car. The rectangular lamps immediately above the bumper are the parking lights and "winking" traffic signals.

conditions short of actually restarting from rest. As a test measure, indeed, this manœuvre can be performed smoothly on top gear, but second is the normal gear for starting on. With the overdrive transmission fitted the top gear or final drive ratio is lower than when the three-speed box alone is used, and at 4.55 to 1 is a comparatively low ratio in conjunction with a large eight-cylinder engine. On this top gear a creep of 5 m.p.h. is a smooth normal possibility, and the car will thread its way in and out of town traffic, including corners, and pick up with no protest whatsoever as the throttle pedal is depressed.

Also, a gradient as severe as 1 in 6½ (nearly 16 per cent) was taken throughout on the ordinary top gear, including even the rounding of a sharp bend at the summit, a performance which has not been equalled by many cars throughout *The Autocar's* experience. It is also notable that at no time, at either low or high speed, when the throttle was opened wide was there any sign of pinking, even on the relatively poor quality Pool petrol supplied in England. The Hudson's compression ratio is lower than in a good many other contemporary U.S. engines.

When the control is in the appropriate position overdrive top gear comes into operation at any speed above 22 m.p.h. upon momentary release of the throttle pedal, and a ratio of 3.18 to 1 is thus introduced. So high is this ratio that there is no particular range of speed that can be quoted as the car's comfortable maximum; it remains easy and effortless on the overdrive gear up to slightly above a genuine 90 m.p.h.

At any speed when the overdrive is in use normal top gear can be regained through the kick-down solenoid-relay control, operated by fully depressing the throttle pedal. To lock the overdrive entirely out of operation may be desirable under certain conditions, as with the overdrive control in the operative position a free wheel functions up to the cutting-in speed of the overdrive and no braking effect from the engine is then available on the overrun. With a quiet, flexible, big engine such as this the presence of a free wheel is scarcely apparent during traffic driving. Synchromesh on second and top gears is effective; the movement into reverse from rest was not always clear cut.

On the Hudson the throttle pedal pressure which has to be overcome to regain normal top gear from the overdrive is lighter than in some other instances. This can be regarded as logical, for as the throttle is depressed to accelerate or for maximum power uphill the ordinary top gear is restored without the driver thinking specially of the process. It does mean, however, that when the ordinary top gear is not actually required one finds oneself obtaining it at times because the point of travel on the throttle pedal where the change-over operates has been reached unintentionally. The change down from overdrive to top gear is not devoid of a slight jerk, especially when performed at speed, but the change up to overdrive is scarcely perceptible beyond a slight clicking sound.

Miles are covered in an effortless fashion which results in average speeds better than 45 m.p.h. being readily achieved, and there is a particular ease about the process to the driver, no small part of which comes from the fact that there is practically no need to touch the gear lever. It should perhaps be emphasized for the benefit of those

A phrase which has been applied to the current Hudson is "the car you step down into," and certainly the floor level is low. In the doors, which carry elbow rests shaped to act as grips for shutting them from inside, are chromium-plated ledges in which the ashtrays are placed conveniently. Nylon loose covers were fitted over the cloth upholstery. In the rear squab is a gigantic folding central arm-rest.

The four-window style now frequently used in U.S. cars is not followed ; the rearward pair of side windows is quite small but serves a useful purpose, being hinged to act as an extractor for additional ventilation control. Within the panel seen in the left rear wing is an electrically locked fuel tank filler. Only very shallow detachable rear wheel shields are fitted.

not familiar with overdrive systems that the changes between normal top and overdrive top are entirely foot controlled, the gear lever remaining in the ordinary top-gear position. As usual, there are also overdrive versions of second and first gears, but these are largely of academic interest.

Although the overall dimensions are considerable the Hudson does not seem from the driving seat to bulk as large as some of its contemporaries, and it does not feel really unwieldy. Driving vision is very fair, although it could be better to the left side. With right-hand controls the left-side wing is just out of sight of an average-height driver, but the right-hand wing is seen. The windscreen pillars are not overpoweringly obstructive, whilst vision over the bonnet, although carrying the driver's sight line farther ahead on the road than is customary in British cars, is not open to major criticism. More disturbing in this respect is the fact, at all events in earlier acquaintance with the car, that the "bay-window" windscreen, of

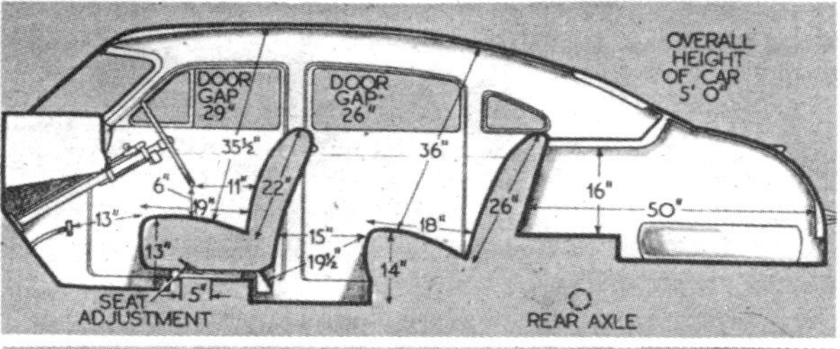

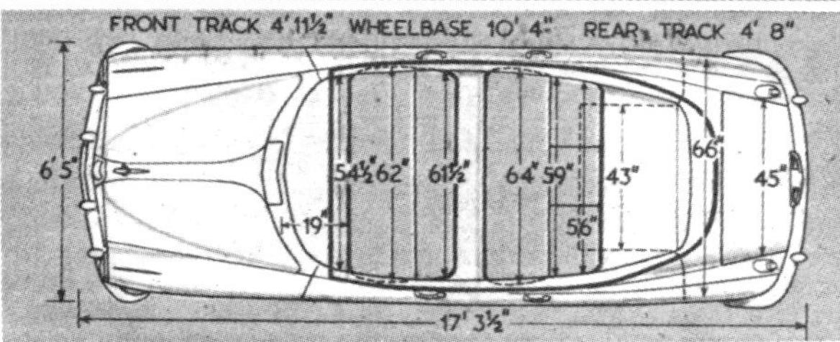

Measurements in these scale body diagrams are taken with the front seats midway between the positions of fore and aft adjustment and with the seat cushions uncompressed.

curved glass in the two individual sections, is a considerable distance away from the driver, which is not ideal in rain at night. Also to one driver at least this windscreen produced curious and at times slightly disturbing reflections.

The driving position as such is excellent, the big-diameter wheel being placed at an extremely comfortable angle where one has the feeling of proper power over it. A point of criticism, in which the Hudson is not alone among U.S. cars, is that the clutch and brake pedals have long arms, resulting in a backwards and upwards movement of both feet to these pedals, which hinders the quickest possible transfer of the right foot from throttle to brake. The brakes are of Bendix pattern with hydraulic operation. There is also a mechanical linkage to the rear brakes, further depression of the pedal bringing this into action as an emergency measure in the event of an hydraulic failure.

In addition to having right-hand controls, and the overdrive transmission, the car tested was fitted with a number of items which rank as extra equipment, the difference in this instance between the basic price in the U.S. and the price of the car equipped as tested being a matter of some $370, or approximately £90.

As expected in an American car, there is ample space for three people on either front or back seat. The latter has an extremely wide central arm-rest for use when only one or two passengers are seated at the back. All doors can be locked from the inside by push-button controls, whilst return to the car can be made by either front door, as both have key locks. From outside the doors are opened by a press-button action incorporated in a handle of conventional appearance.

Efficient Interior Heating

The interior heating system is of the fresh-air type which circulates either cool or warmed air—controllable for temperature—under pressure of the air admitted through a flap vent in the top of the scuttle, which incorporates a rain and snow trap, whilst in addition there is a fan for demisting and de-icing purposes or for increased heating effect at low speeds. On chilly evenings it was found comfortable to drive without top coats with the windows partially open.

Suction-operated screenwipers are driven through wires running over pulleys; these wipers were found to dry up at full throttle. Sealed-beam head lamps give a good beam for speeds in the 70s over known roads. Twin horns are controlled very conveniently by a ring on the steering wheel, and produce a penetrating note. Cupboards at each side of the facia board can be locked. The driving mirror gives a thoroughly comprehensive view, and "winking" traffic signals at front and back are controlled by a lever on the steering column, the action of which is self-cancelling after a turn has been made.

Starting at all times was immediate, an automatic choke operating when the engine is cold. An interesting item of special equipment on this car was an electrically controlled lock for the petrol tank filler. Only when a switch inside the car was depressed, with the ignition switch "on," could the cap be opened, it flying open under spring pressure. Because of the frame design an individual type of hydraulic jack is applied at a special point at each side, and there is a similar socket at the front, though the front wheels, of course, are outside the frame members in the normal way.

The metal facia has a grained finish which closely resembles wood. Radio controls are at the centre, and the speaker grille is in the platform between facia and curved windscreen. Oil and charging indications are by red lights, but an engine thermometer is fitted. The forward outlet of the heater is seen beneath the facia, also the white knob of the temperature regulator ; to the left, on the same level, is the air vent control. The anti-dazzle switch is below the clutch pedal. Above the windscreen, and higher than the radio aerial control knob seen here, is a useful interior light, additional to twin interior lights in the rear quarters.

MOTOR TRIALS

KEEPING APACE WITH THE HUDSON PACEMAKER

by Walter A. Woron, Editor

The car for this road test was furnished to MOTOR TREND *by Jack Gaynor, Hudson car dealer at 1122 S. La Brea, Inglewood, California.*

THE HUDSON Motor Car Company's invasion of the smaller car class shows an apparent trend by many of the large-car companies to get into the lower priced field. The new Hudson Pacemaker, although actually not a small car nor in the low-priced field, nevertheless should give serious competition to the Pontiac-Dodge-Buick Special class.

Anxious to determine how this new car performed, we arranged for a road test with Jack Gaynor, Hudson dealer of Inglewood, California. At 9:25 on the morning of an overcast day, we drove a four-door sedan Pacemaker, painted a beautiful peacock green, out of the garage and headed for the open highway. The speedometer reading at that time was only 172.8 miles, the mileage having been registered in the two days previous to our picking up the car.

The car we used for the test had a conventional transmission, but we soon found that the shifting mechanism was anything but "conventional." The car is very easy to shift, making it particularly adaptable to speed shifting. The combination of helical cut gears and the cork friction surface of the single plate clutch makes clashing of gears almost impossible, unless you accidentally shift from low to reverse.

The shortened hood makes visibility from behind the large steering wheel very good at all times, with no apparent blind spot being evident; however, as with many curved windshield cars, there is a double headlight reflection from oncoming cars at night, but which you become accustomed to.

The steering gear ratio is 18.2 to 1, which makes it very easy to steer and to park. There is the feeling at the wheel that you always have positive control. When a front tire pulls off the pavement onto the shoulder of the road, the car is very easily corrected, with no fighting of the wheel being necessary. Around sharp turns the car has good recovery and does not lay over noticeably. Even with the low pressure tires of 15x7.10 (inflated to their normal pressure of 28-30 psi), there was no noticeable tire squeal. Although the suspension is conventional, apparently the distribution of weight on the front wheels, in combination with the steering gear ratio and the center point steering, makes for easy control.

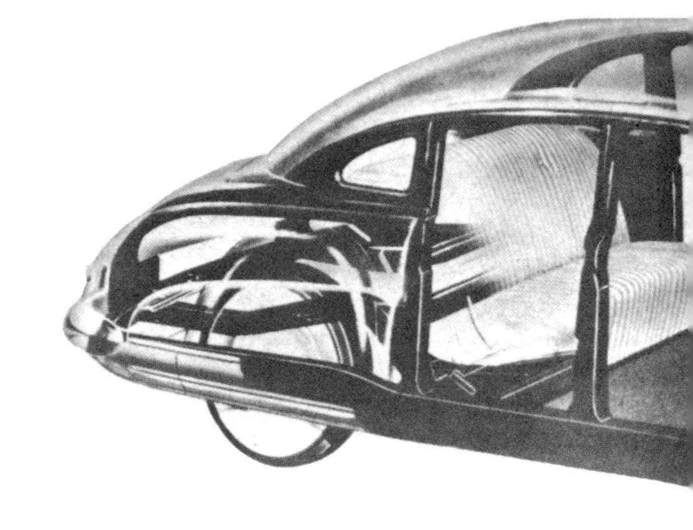

PHANTOM view of Hudson Pacemaker shows Monobilt structure

In testing the car for lugging ability it was found that the car would idle down to eight mph in high and would then accelerate without bucking or jerking. However, there was a small amount of vibration just as the throttle was floor-boarded.

Arriving at our test strip, we set up a quarter-mile course and then began the speedometer check. When the speedometer indicated 30 mph, it was actually doing 28.5. At 60 mph the speedometer was apparently more accurate for we were doing 59.3 mph.

After this calibration, we prepared for the acceleration trials. Results of these tests are shown in the following table:

TEST	TIME (AVERAGE) With Overdrive	W/O Overdrive
Standing Start ¼-mile	:19.83	:20.06
0-30 through gears	:04.79	:04.56
0-60 through gears	:16.45	:15.38
10-60 in high gear	:18.06	:17.95

Two flying runs were made (in opposite directions) through a ¼-mile trap, with the average being 87.73 mph. Since the approach to the ¼-mile course was comparatively short, with a longer approach the car might have attained a higher speed. The speedometer was still climbing at the end of the ¼-mile trap.

During the acceleration trials and after constant use of the brakes, it was noticed that there was a small amount of brake fade; however, the brakes cooled off very rapidly so that in the space of a half mile when the brakes were reapplied, they provided a sure, gradual action. The Hudson brakes are not the kind that tend to throw you on your head the instant you touch them, but instead slow you to a gradual stop.

After the acceleration trials, we drove the car at high cruising speeds through the desert and then began an ascent to an altitude of close to 7,000 feet. At about 6,000 feet we began to run into snow, making a good test of the car through various types of climatic conditions. On one particular test strip of a seven per cent grade, the car in overdrive was able to maintain a speed of 58 mph. When the car was dropped out of overdrive (by accelerating the throttle), the speed picked up to 63 mph.

The fuel consumption check we made was not quite as fair as we would have liked, considering that the car was driven at high speeds, that the acceleration features of the car were being used, and that there was a considerable amount of mountain driving during the test. However, the overall average of 14.5 miles per gallon seems to indicate that the car in normal driving should maintain an average of about 18 to 19 mpg.

Desirable features of this new 119-inch-wheelbase car are manyfold, and although some of them may seem to be small in importance, they show an evidence of many hours of research. The door panels have been cut out so that the window controls and door handles are recessed, providing elbow and body clearance. The doors and the back of the seats are covered with colored Dura-fab, which is a vinyl plastic that reportedly does not scuff, crack, split or peel, is stainproof and can be cleaned with a damp cloth. The sun visors are mounted with a hinge at the center, which should eliminate the vibration and rattle that accompanies the end-mounted type.

As with the Super and Commodore custom series, the Pacemaker has a Monobilt structure, which has a base of eight transverse cross members. These cross members are up to $6\frac{13}{16}$ inches in depth and are joined to four longitudinal members. The outside frame rails are outside the rear wheels. Vertical members extend upwards from the outside rails to the roof. The body and roof panels are then welded directly to these body and frame members. This frame construction, along with the front coil springs and the rear leaf springs, the four airplane-type shock absorbers and low pressure tires, contribute to the very soft ride of the Pacemaker. Stabilizers are used at both the front and the rear, giving added riding stability and helping to eliminate front end sway.

Continued on page **18**

PACEMAKER acceleration qualities are graphically portrayed

PHOTOGRAPHS BY THOMAS J. MEDLEY

THE Pacemaker was also tested in the desert country

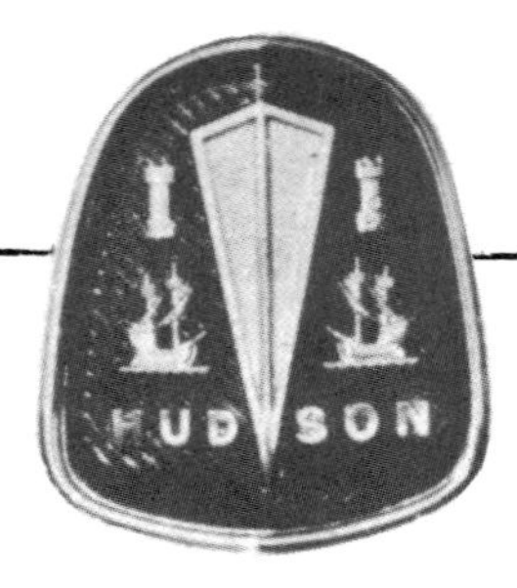

MOTOR TRIALS

HUDSON HORNET FASTEST YET TESTED . . . 97.5 MPH

by Walter A. Woron

PHOTOGRAPHS BY E. RICKMAN

HORNET—if you take this name and Webster's explanation of same literally—is an apt moniker for Hudson's new high-powered six. It hums, buzzes, zooms and practically everything else any self-respecting hornet would do (except fly). And speaking of flying, the top speed of this car isn't far from approaching "take-off" speeds.

The Hudson Hornet is impatient—it wants to go, and runs best fast. It's powerful, is fast in the getaway, is easy to handle and gives a comfortable ride. There's not much more that can be asked of a car, except for personal tastes, price and upkeep cost.

The test car for this motor trial, a Hornet four-door sedan with Hydra-Matic transmission, was made available to MOTOR TREND Research through the Hudson Motor Car Company's Los Angeles Zone Office. The factory car, which was fairly well broken in (1332 miles on the odometer) was picked up for the two-day test from Wright & Beal, Hudson dealers at 1060 So. Figueroa St., Los Angeles, Calif.

Detailed Test Report

ACCELERATION: Although the Hornet is not a light car (3880 lbs.), the acceleration characteristics of the car are more than favorable. In fact, in the standing start ¼-mile department, it can give competition to the best of them. In all instances, except from 0 to 30 mph, we were able to improve acceleration times using the LOW range of the Hydra-Matic transmission, then shifting to DRIVE. In the standing start ¼ mile, we wound the car up to around 40 mph in LOW, then quickly shifted to DRIVE. This seemed to be the best shift point and gave the fastest elapsed time. Using this method we were clocking 70 mph at the end of the quarter mile.

Oddly, in the 0 to 30 mph checks, we found the DRIVE range to be faster than the LOW range. At these lower speeds, in LOW or DRIVE there is at least one shift (first to second gear in each case), but the shifting time seemed to be faster in the DRIVE range. The good acceleration time in the 30 to 60 mph check resulted from the fact that when the throttle was tromped on hard, the transmission downshifted to third. The same was true of the 10 to 60 mph tests, where the transmission downshifted to second gear before acceleration could begin.

TOP SPEED: Not much need be said of a car that can turn an average top speed of 97.09 mph, which is the highest average speed yet turned by MOTOR TREND Research during a motor trial of any stock American production car. The Hornet is unusually stable at its top speed. On the car's fastest run, a clocked 97.51 mph, the speedometer was indicating 106 mph, which is not too great a speedometer error at these high speeds. (For speedometer correction, see Table of Performance.)

BEHIND THE WHEEL: Here you first get the impression that you're in a large car and you wonder if it's going to be hard to drive. But, as you step on the throttle and begin to move through traffic, you find that it is quite easy to maneuver and to judge distances (even though you cannot see the right front fender from the driver's position). There's plenty of leg- and headroom (38¾ ins. in front) and all operating controls are handy.

Two disconcerting features, however, one which was immediately noticeable and another which became evident after considerable driving on the open highway, are the speedometer location and the rear window distortion seen in the rear view mirror. The speedometer is on the far left of the instrument panel and is hard to see with your left hand on the upper section of the steering

Continued on page **18**

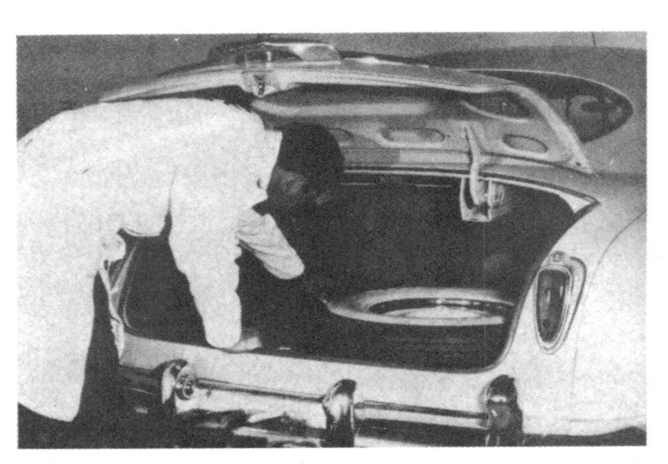

TRUNK compartment, though spacious and deep, does not have much height. Spare wheel takes up considerable room in this position, but it and bumper jack are accessible, as demonstrated by Don Francisco

HUDSON Hornet engine is entirely new six, with more rigid and thicker cylinder walls. Engine is similar to other Hudson engines except that it has larger bore and longer stroke. The bottom end is also beefed up

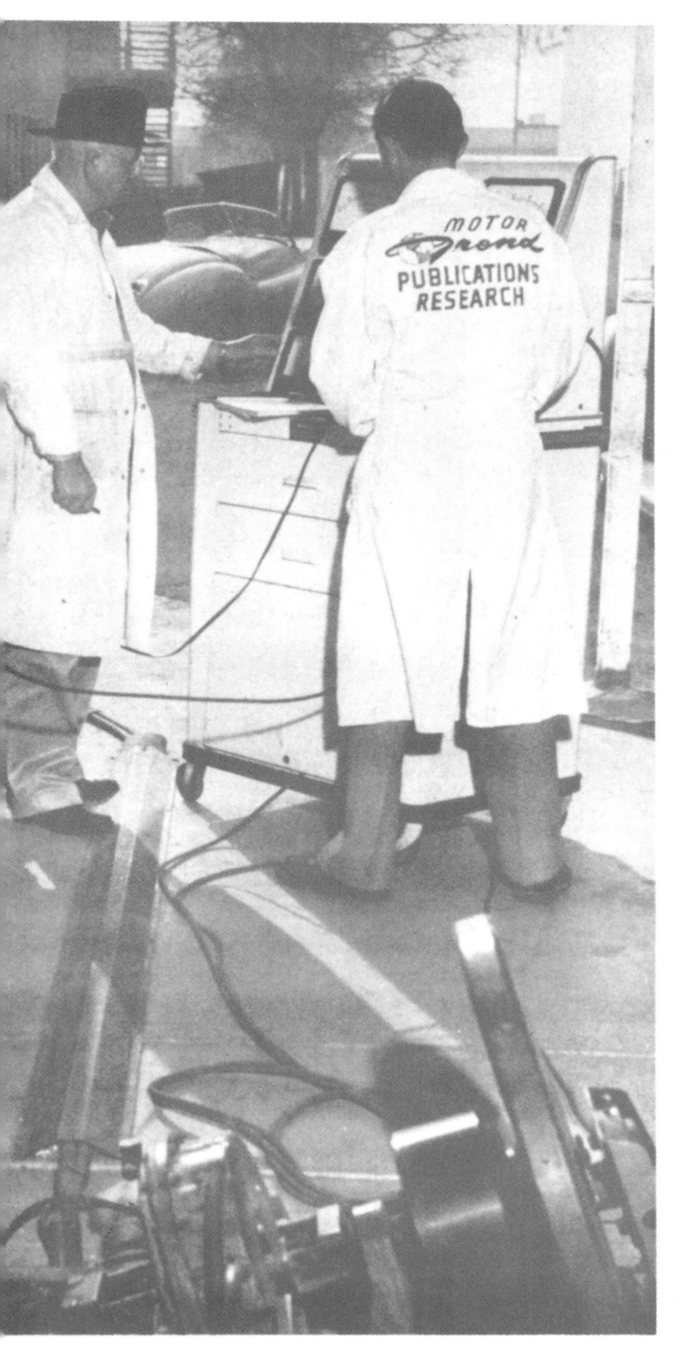

IN DYNAMOMETER check at Clayton Mfg. Co., car delivered maximum of 90 road hp @ 3300 rpm @ 74 mph. This figure is net delivered hp at rear wheels, taking into account normal losses through transmission, differential and rolling resistance. Dynamometer unit is in foreground

OVER DIRT, or bumpy asphalt roads, ride is comfortable. Car is very steady into turns, with only a small amount of body heel. Over dips at high speeds, car does not bottom. Not necessary to fight wheel on car tracks or on asphalt ridges. Hydra-Matic not at its best in dirt

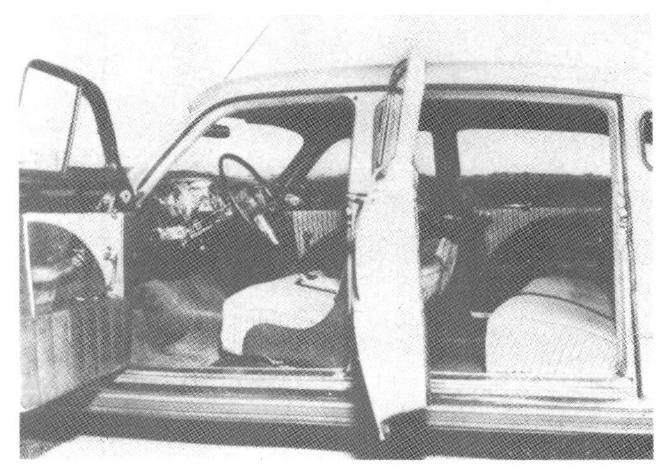

LUXURIOUS interior includes nylon and Dura-fab upholstery, foam rubber seats, well-made appointments. Dash has considerable chrome, but not in location to cause glare. Top of dash now covered with plastic. Visors are well constructed and are adjustable to all positions

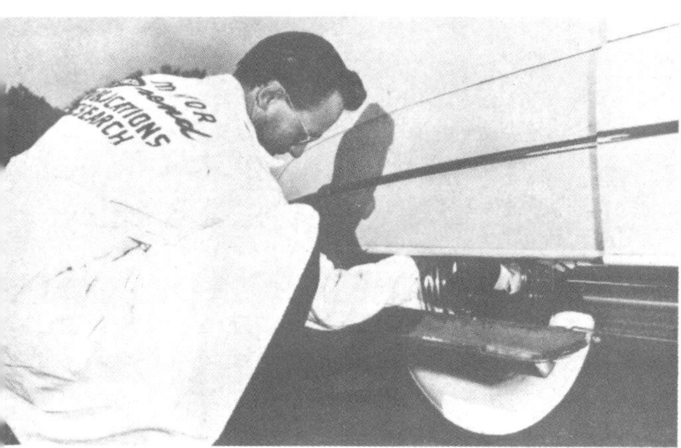

CHECKING rear tires for air or changing them could be a problem, except that rear skirt has been designed to fold down and can also be easily removed. Skirt is designed as a part of the fender panel

BRAKES of Hornet proved to be adequate, with stopping distances at high speeds being amazingly short. Hudson brakes are hydraulic, with reserve mechanical system to take over in case of hydraulic failure

Hudson Motor Trials

Continued from page **16**

wheel. The rear window makes cars appear squatter than they actually are—and by raising or lowering yourself in the seat, you can cause them to change their shape.

STEERING: Despite the high steering gear ratio (20.4:1) the Hornet handles as if it had a much quicker ratio. Controllability is good into turns at high speeds and it doesn't take a lot of steering wheel movement to keep the car going around a corner. Center-point steering is used in conjunction with the 18-in. steering wheel, which is set at a comfortable position. No wind wander was noticed at any time and, surprisingly, we only heard the low-pressure (15 x 7.10) tires squeal under the most severe cornering.

TRANSMISSION: The Hudson Hornet comes stock-equipped with a conventional three-speed forward transmission, to which overdrive can be added ($99.92 extra). The test car, however, was equipped with Hydra-Matic (optional, $158.30), with which a 3.58:1 rear axle is used.

FUEL CONSUMPTION: The over-all fuel consumption average for the Hornet on this motor trial was 14.07 mpg, which is not overly high; however, in consideration of the car's weight and the fact that a big displacement (308 cu. in.) engine needs a lot of fuel to keep it running, this average is not too disappointing. The averages for various speeds and traffic conditions are shown in the Table of Performance.

Appearance and Mechanics

If you've ever seen a Hudson after it has been involved in an accident (or have seen photographs of one), you've seen the greatest advantage to the unit-type, all-welded frame and body, such as used on the Hornet. The Monobuilt frame completely encircles the passengers, front, sides and rear, while the floor is recessed below the frame. The safety aspect of this arrangement cannot be denied, while the merits of the "step-down" floor are dependent on personal likes and human configurations.

The body of the Hornet seems to be well put together, the welds and fit of joints being of good workmanship. Body styles available in the Hornet series include the club coupe, convertible brougham, the four-door sedan.

Hudson claims to have the hardest engine block in the industry—a chrome alloy in which no valve seat inserts are needed. The most powerful six used in American stock production cars, the engine is a continuation of Hudson's policy of developing high compression L-head engines.

Some interesting features of the Hornet engine are: a dynamically and statically balanced, four main bearing crankshaft; pinned-in-position piston rings (to prevent chatter and rotation); pressure lubrication and a floating oil intake (to permit recirculation of the cleanest oil); a dual carburetor (three cylinders each being fed by a separate carburetor system); and a waterproof ignition system that has a radio and television suppressor built into the distributor cap. (If you have a TV set, you know what a boon this would be as a requirement for all cars.) Detail specifications are given in the Table of General Specifications.

Trend Trials Number

As we have done with each tested car, starting with the Nash Statesman (Dec. '50 issue) we have determined a Trend Trials No. for the Hudson Hornet. This number has been arrived at through totaling the cost per bhp (how much it costs you for the power of the automobile), the fuel cost per year (based on the over-all fuel consumption average obtained on the motor trial), and the normal maintenance cost.

Ordinarily, the Hudson Hornet with Hydra-Matic would have been given a Trend Trials No. 29.8, the lowest number given to a car evaluated in this price bracket ($2201-$2500). However, there is another very important item that we are adding to the TT No. this month—that of depreciation, something which cannot be overlooked.

The average depreciation percentage is arrived at as follows: The average retail price of the models from the five previous years is each subtracted from the new model's price. When this is averaged, it gives the average depreciation percentage, projected over a one-to-five-year period. (This should *not* be construed as the *average yearly* depreciation.)

In the case of the Hornet, the depreciation percentage is 9.9. When this is added to the above-quoted TT No. 29.8, it gives a final Trend Trials No. of 39.7 for the Hudson Hornet with Hydra-Matic.

(Since the depreciation percentages were not worked out for the three previous cars on which TT Nos. were quoted, they are given here: Nash Statesman with Hydra-Matic, 8.8, for a revised TT No. 32.9; Ford with Fordomatic, 8.1, for a revised TT No. 30.2; and Packard with Ultramatic, 11.3, for a revised TT No. 43.7.)

TABLE OF PERFORMANCE

DYNAMOMETER TEST	
1200 rpm (full load) 25 mph	38.5 road hp
1800 rpm (full load) 39 mph	58 road hp
2000 rpm (full load) 43 mph	63.5 road hp
3300 rpm (full load) 74 mph	90 road hp (max)

ACCELERATION TRIALS (SECONDS)	
Standing start ¼-mile	:19.88 (D)*; 19.41 (L-D)**
0-30 mph	:04.41 (D); 05.11 (L)
0-60 mph through gears	:14.95 (D); 14.64 (L-D)
10-60 mph in high	:13.74
30-60 mph in high	:10.39

*Shift using DRIVE range only. **Shift using LOW range, then to DRIVE.

TOP SPEED (MPH)	
Fastest one-way run	97.51
Average of four runs	97.09

FUEL CONSUMPTION (MPG)	
At a steady 30 mph	18.50
At a steady 45 mph	16.91
At a steady 60 mph	14.28
Through light traffic	13.62
Through medium traffic	11.49
Through heavy traffic	9.60

BRAKE CHECK	
Stopping distance at 30 mph	30'4"
Stopping distance at 45 mph	92'8"
Stopping distance at 60 mph	166'0"

DECELERATION CHECK (SECONDS)	
Time to slow from 60 to 45 mph	:09.61
Time to slow from 45 to 30 mph	:11.81
Time to slow from 30 to 15 mph	:16.28

GENERAL SPECIFICATIONS

ENGINE	
Type	L-head, in-line 6
Bore and Stroke	3 13/16x4½
Stroke/Bore Ratio	1.08:1
Cubic Inch Displacement	308
Maximum Bhp	145
Compression Ratio	7.2:1

DRIVE SYSTEM

Transmission—Conventional three speed; Ratios: Low—2.88:1; Second—1.82:1; Third—1:1; Overdrive—.7:1; Hydra-Matic four-speed, automatic transmission

Rear Axle—Hypoid, semi-floating. Ratios: Standard—4.10:1; Optional—3.58:1 or 4:55:1; with Hydra-Matic—3.58:1; with overdrive—4.55:1, optional 4.10:1

DIMENSIONS	
Wheelbase	124 ins.
Overall Length	208 3/32 ins.
Overall Height	60⅜ ins.
Overall Width	77 2/32 ins.
Road Clearance	8 ins.
Tread	Front, 58½ ins.; Rear, 55½ ins.
Turning Radius, Right, 21 ft. 2 ins.; Left, 20 ft. 5 ins.	
Turns, Lock to Lock	5⅞
Weight (Test Car)	3880 lbs.
Weight/Bhp Ratio	26.8:1
Weight/Road Hp Ratio	43.1:1
Weight Distribution (Front to Rear)	56.4/43.6%

MOTOR TRIALS . . . the Hudson Pacemaker

Continued from page **15**

The powerplant of the Pacemaker is a 112-horsepower, six-cylinder engine that is pressure lubricated, unlike earlier Hudson engines. This engine is actually the late '48 Super engine, with the same bore ($3\frac{9}{16}$ inches) and a shorter stroke ($3\frac{7}{8}$ inches), bringing it close to a "square" engine. The engine has a displacement of 232 cubic inches, and has a standard compression ratio of 6.7:1 that can be increased to 7.2:1 with an aluminum head. All component parts of Hudson engines are dynamically balanced, including the crankshaft, the clutch, the flywheel, the connecting rods, and the pistons.

Moving into the low medium-priced field with the Pacemaker seems to have been a wise move by the Hudson Motor Car Company, for according to its performance figures, the car should provide healthy competition in that field. And for a person who is willing to pay slightly more than the price of a low-priced car for additional comfort, the Hudson should make a very satisfactory buy.

HUDSON HORNET

FROM THE WOMEN'S ANGLE

Just take a good look at that powerful 145 H.P. Hudson Hornet engine—it'll do a hundred easy, but 75 is plenty fast for me. I like the wide doors, makes getting in and out easy as pie!

"CANDY" JONES, POPULAR MODEL AND TV STAR, GIVES AUTO SPORTS REVIEW READERS HER OPINION OF HER NEW HUDSON "HORNET"

SEE NEXT PAGE

HUDSON HORNET FROM THE WOMEN'S ANGLE

There's plenty of room inside my Hudson, I'm never crowded even with five or six in the car.

The dash board is smart, compactly arranged instrument panel makes for easy "eye-level" reading.

High compression engine—6-cylinder, L-head design; 145 horsepower with 7.2 to 1 compression. Miracle-Dome aluminum head. Bore 3 13/16 inches; Stroke 4½ inches. Piston displacement 308 cu. in.

I'M no expert on the internal combustion engine, but I, like most other women, have my own reasons for choosing a certain toothpaste, cigarette or car. Now take my new Hudson "Hornet." It's got everything anybody would want in a car.

I like a convertible, I enjoy the outdoors,It's got a top that goes up and down automatically, all I do is press a little button! My "Hornet" has lots of "get-up-and-go," I can pass any car on the highway! It's safe, has good brakes, is easy to stop and a cinch to park.

Photography by
M. P. Goldschmidt ARPS

I find that the Hudson "Hornet" is easy to park, easy to steer, even over rough cobblestone roads.

I like lots of luggage room in a car. Here is Edith Glade and two hat boxes to prove my point.

The car is spacious, has lots of leg room. I've had as many as nine adults in it at one time! I can stash away five average size over-night bags in the spacious trunk compartment. The seats are tastefully upholstered in red leather, as are the sides and doors. I find the chrome-plated appointments both functional and smart in design.

The steering wheel is positioned just right, as are the foot and hand controls. All instruments are grouped directly in front of me, I like the aviation type speedo and clock. The front seat is adjustable and the windows operate automatically. A remote control switch at the driver's left also operates all four windows. Interior and exterior finish, right down to the soft rug, gives that custom-made appearance usually found only in the most expensive European models.

My Hudson is low, long and has that modern, rakish look, I feel absolutely safe and secure when I step on the throttle and whiz by car after car on the open highway. I've had rides in more expensive makes but I can't think of any car I'd like any better, than my Hudson "Hornet."

SEE NEXT PAGE

HUDSON HORNET FROM THE WOMEN'S ANGLE

Hudson's "step-down" design makes it one of the lowest and smartest cars on the American highways today.

★

The top of my Hudson folds away in a jiffy—automatically. All I do is press a button—it works like magic.

★

Thanks to the Hudson-engineered 145 horsepower engine I can take the steepest grades with ease!

HUDSON

There isn't a better built automobile in America today than the Hudson Hornet, regardless of price. As to roadability, the Hudson is head and shoulders over any other car made in the U.S.A. It is America's safest car because it is our best road car.

Hudson Hornets have been cleaning the slate in just about every important stock car race held in the United States. Not because they are faster—they are not—but because they out-handle all our other cars.

The Hudson Hornets can be bought stock or with factory hop-up equipment that makes them real bonafide wild cats. I recently tested a Teaguemobile (Hudson Hornet assembled and tuned by race driver Marshal Teague) on the Le Mans course. This car, owned by Briggs Cunningham, had the same engine used by Teague in finishing 6th in the 1951 Mexican Road Race. The acceleration was only fair for such a hop-up (zero to 60 in 12.1 seconds) but the top speed was a real 107 m.p.h. My test took place just two days before the 24-hour race and I found I was going into corners and getting out of them better than many of the practicing race cars. It had export shocks and springs, 8 to 1 compression ratio, and two carburetors, all factory optional equipment. The car is extremely comfortable and all appointments are top grade. It is a Jumbo for room, in fact for my tastes, actually too big. The styling does not appeal to me but everything else does. The brakes are far superior to average and the steering is so much better that comparison is silly.

The Hudson Hornet is one of America's great cars, and may very well be its greatest at the moment. Hudson also makes the smaller engined Pacemaker and, by the time you read this, may have introduced its new small car, being built to compete with the Nash Rambler and Henry J.

specifications:

6-cylinder, L-head, 308 cu. in, displacement; 145 h.p. at 3,800 r.p.m.; torque, 257 foot pounds at 1,800 r.p.m.; bore, 3.81″, stroke 4.50″; 124″ wheelbase; over-all length 208″; width 77⅝″; height 60⅜″; tire size 15x7.10; 20 gallon tank. Performance: 0 to 30, 4.2 seconds; 0 to 50, 9.2 seconds; 0 to 60, 13.6 seconds; 0 to 70, 18.9 seconds; top speed 97 to 98.

ROAD and TRACK ROAD TEST No. A-4-52

Hudson Hornet

SPECIFICATIONS

6 cylinder in-line L-head engine		Weight as tested	3840 lbs.
Compression ratio	7.2 : 1	Weight, front	2230 lbs.
Horsepower	145 at 3800 rpm	Weight, rear	1630 lbs.
Torque ft/lbs.	257 at 1800 rpm	Overall length	201.5 in.
Bore and stroke	3.81 x 4.50	Overall width	77.12 in.
Displacement	308 cu in.	Seat width	64 in.
Transmission	Hydramatic	Seating cap.	3 front, 3 rear

Total price (delivered in Los Angeles) $3650

TAPLEY READINGS

Pulling power	187 lbs per ton
Rolling resistance—	
10 mph	34 lb per ton
60 mph	68 lb per ton
Braking efficiency	87%

PERFORMANCE

Flying ¼ mile	92.9 mph
Fastest one way	97.8 mph
Standing ¼ mile	20.7 secs.
Lo-range, 32% grade	21 mph
Hi-range, 32% grade	19 mph

FUEL CONSUMPTION

Steady	Miles per gal.
30 mph	15
40 mph	13
50 mph	12.5
60 mph	12
70 mph	11.5
80 mph	11
City traffic	9.5

SPEEDOMETER CORRECTION

Speedometer	Actual
30 mph	29.6
40 mph	38.6
50 mph	47.9
60 mph	57.3
70 mph	66.6
80 mph	76.9
90 mph	87.3
98 mph	97.8

JOHN BOND REPORTS

John's Automotive Biography

Cars owned: '28 Ford, '29 Ford, '30 Ford, '35 Chevrolet, '37 Buick 60, '40 Chevrolet, '39 Buick '40, '33 Terraplane 8, '46 Ford 6, '32 Ford V-8, '47 Ford V-8, '49 Ford V-8, '49 MG TC. Now owns: '36 Ford, '51 Chevrolet, '50 Ford sports car under construction. Doesn't usually write up road tests because he's too critical. Likes smaller cars, but not too small. Would like to own a Willys Aero for family use, a Ferrari America open 2-seater, and a 350 cc Jawa motorcycle.

An opportunity for taking the wheel of the Stock Car Champion was eagerly accepted, and in all, I drove it about 250 miles.

Looking over the exterior of the car . . . it is obviously bulky, yet low . . . and the bumblebees plastered at various vantage points leave no doubt that this is a Hornet. Chrome seems to swirl about with little rhyme or reason, and a scraped spot on one side gave evidence that the heavy L. A. traffic could require better-than-average skill with this bigger-than-average "boat".

The interior is impressive, tho I felt lost behind the wheel. The windshield is too far away, the instrument markings are flamboyant, and there's no convenient spot for the left elbow. This is not a car that you can jump into, drive off, and feel at home in—it takes some learning to judge your nearness to others on street and highway.

The Hydramatic indicator might as well have been omitted . . . the markings didn't agree with the detents. But I finally found drive-range and moved off.

The huge 308 cu in. 6-cylinder engine is very smooth at all speeds, and quiet during normal driving. Under full throttle it produces a power roar (which becomes less noticeable at high speeds when wind noise covers the roar). Wind noise, however, is not excessive . . . the air resistance of this body must be low. Passengers in the car liked this power roar and it did make the performance seem all the more impressive. Performance is definitely there, as the acceleration figures prove. The speedometer readings could only be estimated within 5

Tho it provides a handy shelf in drive-ins, the forward placement of the Hudson Hornet's windshield cuts down visibility in heavy fog.

mph, because of the thickness of the needle and dial markings, but the speedo seemed unusually accurate. A reading of approximately 95 mph took a long stretch of wide-open throttle, and 100 appeared only on a long downgrade.

Handling qualities are undoubtedly a good part of the success story of the Hornet's stock car career. *Road and Track's* test model did not have the optional equipment (ingeniously called the "export kit") which includes reinforced wheels and hubs, stiffer springs and heavy duty shocks. However, it does handle better than other American cars of the same size and weight. The steering is very slow but it is not spongey at low speeds. High speed turns gave a rubbery feel that left me uneasy, but I think this impression would be overcome when one gets completely familiar with the car. I felt the steering was more sensitive to road camber variation than is desirable and an hour's run on a slightly cambered super-highway proved tiresome due to the constant pull in one direction. I also thot the caster return in city driving was rather vicious. General riding qualities have always been a top Hudson feature and were excellent on this car. With 32 psi in the ELP tires, there seemed less tire squeal then with most cars so equipped.

The brakes were down slightly but seemed satisfactory. I like the safety feature of a mechanical follow-up on the hydraulic brakes, even tho it is operative only on the rear shoes and ties in with the hand-brake cables. I've always wondered why they even bother with the hydraulic lines to the rear. Many foreign cars use hydraulic brakes front and mechanical rear—an arrangement

Hornet buyers, interested in performance, may not worry about high fuel consumption figures.

Chronic complaint among owners is difficulty of egress when parked alongside another car.

I like very much. On a clutchless car I think the brake pedal should be much larger than the Hornet's—to facilitate use by the left foot.

The optional Hydramatic transmission got a lot of special attention. In the first place, I hate all automatic transmissions, but if I had to buy one, it would only be Hydramatic. This unit has 4-speeds and no more power losses than a conventional transmission. The two top ratios are especially useful, 3rd as a performance gear, and 4th as an economical cruising gear (3.58:1). Unfortunately, as supplied by GM to Hudson, there are a few drawbacks to this transmission. First, it was not the latest type and suffers from not having the dual range feature or a genuine economy axle. The car starts in first gear, but shifts too soon into second. The shifts were definitely three times as bumpy as the new Olds 88 (which I tried out for comparison). The Hudson shifts from 3rd to 4th at 65 mph under full throttle, and this shift was smooth.

If you slow down to about 5 mph and then suddenly tromp on the gas, the car takes a terrific lurch due to a late shift into first gear. This can and does wipe out rear axles very quickly, and not just on Hudson. If you apply full throttle from a standstill, the acceleration is good, all the way up, but back off to 20 or 30 mph and try it again—it's flat, no go! What happens is that from a standstill each gear works in turn to what the engineers call the optimum shift point. But if you're cruising in 4th gear at say 25 mph and apply full throttle, you do not go back to 2nd gear, (which gear you would be in at 25 mph, when starting from zero). Instead, you downshift to 3rd gear (5:1) and the engine is not speeded up enough to give the best performance.

The Hudson is not my dream car by any means, but if your wife wants a bigger car than the Jones' and you want performance, durability, and better than average roadability, the Hudson Hornet is the car for you.

OLIVER BILLINGSLEY

Because I lean toward the small imported car, while John Bond generally favors the domestic product, our opinions of automobiles are usually at wide variance. It was a real surprise, therefore, to find our Hudson Hornet reports almost identical.

Part of my test run was made on the Pacific Coast Highway on a very foggy night. As the white goo swirled about the arm-length windshield, visibility sank toward zero. If, on a foggy day, you will gaze out of a window while standing close to the pane, you will find that the visibility is much better than that obtained by standing some distance back. By the same token, a windshield and steering wheel design which allows the driver to "mug" the glass provides maximum visibility.

The body contours of the entire Hudson line were developed by Hudson stylist Frank Springer and consultant Reid Railton to give both streamlining and high strength. Altho the Hudson will never be considered a "classic" design . . . too much chrome and stuff . . . it is one of the cleanest designs on the market. I particularly like the absence of "rocket-jet-spaceship" influence and the conservative rear fender treatment.

Hornet was as water tight as any car so far tested and the design of the body seemed to keep wind noise down to a pleasing minimum.

RESEARCH ROAD TESTS

TWO HUDSONS

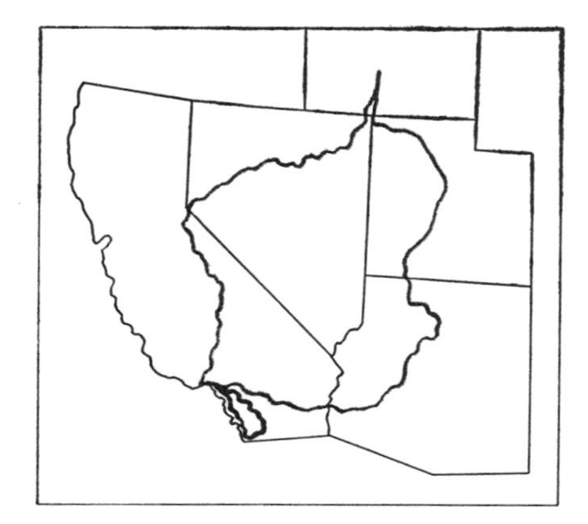

TEST COURSE
—— HORNET
—— WASP

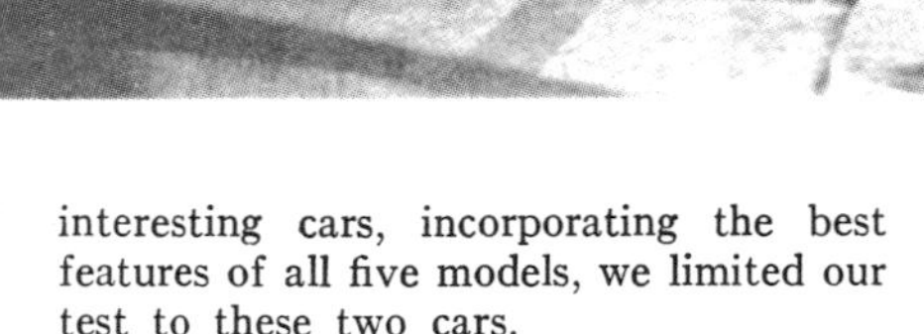

By WALT WORON

FAST COMFORTABLE and safe—a brief but positive description of the Hudson Hornet and Hudson Wasp that can be made without any fear of contradiction. Both cars have top speeds close to the 100 mph mark; they give you a comfortable ride; and with their low center of gravity, responsive steering, and good brakes they are as safe, if not safer than most cars on the road today.

MOTOR TREND Research became acquainted with the Hudson Hornet back in the early part of 1951; at that time we had the same comments to make about the car. Being acquainted with the car, we were more than pleased when the Hudson Motor Car Company turned over one of their demonstration Hudsons to us for use on the 1952 Mobilgas Run. We needed a car that was fast and responsive because we had to play leapfrog with all 26 cars on the Run (spread out over the length of several miles). Of the 1415 miles from Los Angeles to Sun Valley, from below sea level to 8010 feet, we had to continually pass all cars in the Run, stop in time to shoot photographs at various points, start up again and repass them for the same procedure later. For these conditions the Hornet was an apt choice.

Our Test Cars

Hudson builds two basic chassis (one on a 119-inch wheelbase, the other on a 124-inch wheelbase) and four engines 145 hp 6, 128 hp 8, 127 hp 6, 112 hp 6). The combination of these chassis and engines results in five distinct models—Hornet (145 hp 6, 124-inch wheelbase); Commodore 8 (128 hp 8, 124-inch wheelbase); Commodore 6 (127 hp 6, 124-inch wheelbase); Wasp (127 hp 6, 119-inch wheelbase); and Pacemaker (112 hp 6, 119-inch wheelbase). Due to the difference in engines, weight and wheelbases, there are distinct differences between all cars in the line: one is faster, another is more economical, another is more comfortable. However, since the Hornet and Wasp are (on paper, and in actuality) the two most interesting cars, incorporating the best features of all five models, we limited our test to these two cars.

Both the Hornet and the Wasp have better than average acceleration, the Hornet (with its larger powerplant) being the faster of the two through all speed ranges. This powerful six has a lot of punch—always enough for any situation. It has terrific torque at all speed ranges, from 0 up through 70 and 80 mph.

There is no discernible difference between the various gear combinations when trying for best acceleration. It seemed that no matter where (at what speed) a shift was made from LOW to DRIVE, we could not improve on the figures obtained while using DRIVE alone, indicating that Hudson has done considerable testing to determine the best shift points.

Each of the cars are in the 100 mph bracket; the Hornet has a top speed of 99.2 mph (average of four runs) and the Wasp is only 1.4 mph slower. At the top range of speed each of the cars has a certain amount of "floating" sensation, with the Wasp almost giving you the sensation of being airborne. At anything up to top speed, however, both cars were very stable.

Walt Woron records performance data on Hudson while on Economy Run trip

Checking air pressure on the rear tires made difficult by extremely low skirts

Luggage compartment size is ample. Carpeting is gray & green woven fabric

The HORNET and the WASP are above-par automobiles. With low center of gravity, good brakes and steering, Hudson cars are safe

PHOTOS BY JACK CAMPBELL

HORNET: This is one of the best cars for all-around handling qualities that Motor Trend Research has tested. Although the steering is a little slow, control is very good. There does not appear to be too much weight on the front end, which makes it a satisfying car to drive fast on open stretches of highway and over mountainous grades. The only time that we noticed any mushiness was at full throttle going through a turn.

One objection that we found on our test car was some wheel tramp over 70 mph. This may have been due to the fact that the wheels were not properly balanced. On a perfectly straight highway at high speeds, a certain amount of wheel correction was necessary, which was also the case in wind and on wet pavements.

WASP: Through mountainous, curving grades, this car handles sweetly. We found that you could really throw it into turns and seldom, if ever, find yourself getting into trouble. The body wouldn't heel excessively and if you found the rear end sliding a bit, a slight amount of wheel correction into the slide forestalled any possible trouble. Traveling along railroad tracks we found that the body did not whip from side to side but gently *eased* sideways. The one objection was that on a road surface that had been grooved to provide a better traction surface, constant wheel correction was necessary, since the car tended to weave from side to side.

Ride's Comfortable, Stable

HORNET: Probably because of its wheelbase and the suspension set-up, the Hornet literally floats over bumps, rough surfaces and railroad tracks without any uneasy side-sway or "walking." At all times it felt very comfortable and stable. The car has exceptional shocks that grab hold quickly after you hit a dip. The ride in the back seat is not as comfortable as the front seat because there is more side-sway and because of the psychological disadvantage of not being able to see forward over the dash and hood.

WASP: This car rides considerably softer than the Hornet—it has more float, more bounce and more sway. The suspension is, undoubtedly, set up softer to compensate for the shorter wheelbase. Road rumble

(*Continued on next page*)

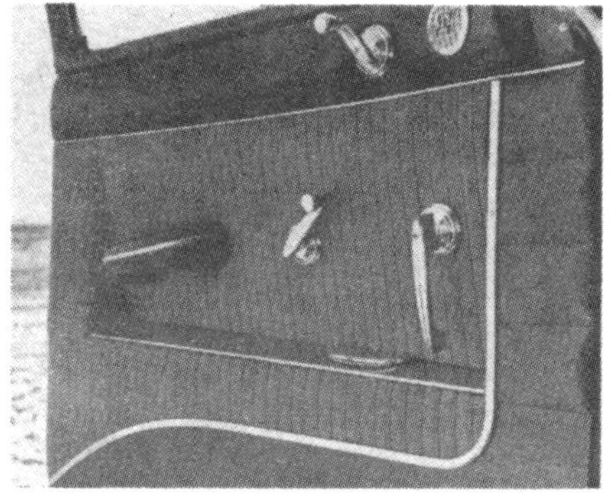

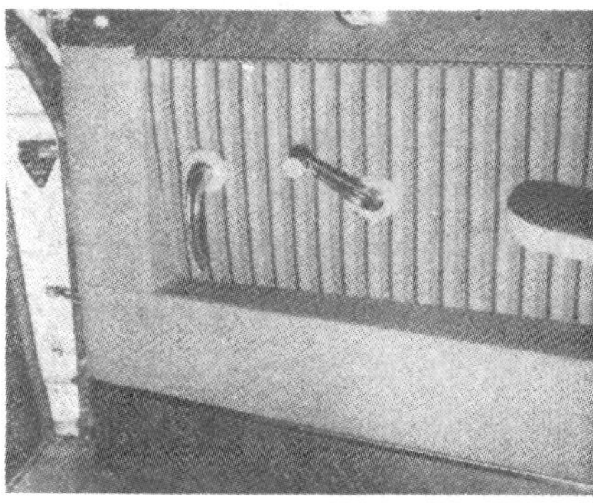

Hudson has reversed usual position of the window crank and the door handle

Interior of Hornet is particularly plush, but over-chromed dash is annoying

Cumbersome method of the Wasp hood release is shown by Tech. Ed. Bodley

Light above windshield gives good interior illumination, is door-operated

HUDSON HORNET TEST TABLE

PERFORMANCE

CLAYTON CHASSIS DYNAMOMETER TEST
(All tests are made under full load conditions)

RPM	MPH	ROAD HP
1200	31	38
2000	54	60
2950	79 (maximum)	73.5

Per cent of advertised hp delivered to driving wheels—50.7

ACCELERATION IN SECONDS
(Checked with fifth wheel and electric speedometer)

	LO-D3	D-3	D-4
Standing start ¼ mile	:20.2	:20.2	:20.2
0-30 mph (0-33, car speedometer reading)	:05.3	:05.2	—
0-60 mph (0-67, car speedometer reading)	:16.8	:17.3	—
10-60 mph in DRIVE	—	:15.6	:18.1
30-60 mph in DRIVE	—	:12.1	:17.3

TOP SPEED (MPH)
(Clocked speeds over surveyed ¼ mile)

Fastest one-way run	105.0
Average of four runs	99.2

FUEL CONSUMPTION IN MILES PER GALLON
(Checked with fuel flowmeter fifth wheel, and electric speedometer)

	D-3	D-4
Steady 30 mph	14.7	21.3
Steady 45 mph	14.2	18.0
Steady 60 mph	12.8	16.2
Approximate average in traffic	—	13.1

BRAKE STOPPING DISTANCE
(Checked with electrically actuated detonator)

Stopping distance at:	
30 mph	46 ft. 1 in.
45 mph	110 ft. 0 in.
60 mph	203 ft. 2 in.

SPEEDOMETER CHECK
(Checked with fifth wheel and electric speedometer)

ACTUAL	INDICATED	ERROR (%)
30	33	10
45	51	13
60	67	12

Odometer correction factor for 100 miles 0.6

GENERAL SPECIFICATIONS

ENGINE

Type	L-head
Bore and stroke	3$\frac{13}{16}$×4½ in.
Stroke/bore ratio	1.18:1
Compression ratio	7.2:1
Displacement	308 cu. in.
Advertised bhp	145 @ 3800 rpm
Bhp per cu. in.	0.471
Maximum torque	257 lbs. ft. @ 1800 rpm
Maximum bmep	125 psi

DRIVE SYSTEM

Transmission—Standard: Silent synchronized mesh, blocker type with helical gears. Optional: Hydra-Matic fully automatic drive; overdrive with standard transmission.

Rear axle: Semi-floating with alloy hypoid gears and alloy axle shafts.

DIMENSIONS

Wheelbase	124 in.
Tread	Front—58½, Rear—55½ in.
Wheelbase/tread ratio	2.1:1
Overall width	77¼ in.
Overall length	208$\frac{15}{32}$ in.
Overall height	60⅜ in.
Turning radius	20 ft. 5 in., left; 21 ft. 2 in., right
Turns, lock to lock	5¾
Weight (test car)	3960 lb.
Weight/bhp ratio	27.3:1
Weight/road hp ratio	54.0:1
Weight distribution (front to rear)	45-55%

INTERIOR SAFETY CHECK CHART

QUESTION	YES	NO
1. Blind spot at left windshield post at a minimum?		X
2. Vision to right rear satisfactory?	X	
3. Positive lock to prevent doors from being opened from inside?	X	
4. Does adjustable front seat lock securely in place?		X
5. Minimum of projections on dash-board face?	X	
6. Is emergency brake an emergency brake and is it accessible to both driver and passenger?		X
7. Are cigarette lighter and ash tray located conveniently for driver?	X	
8. Is rear vision mirror positioned so as not to cause blind spot for driver?	X	
Total for Hudson Hornet and Wasp		62.5

OPERATING COST PER MILE ANALYSIS

1. Cost of gasoline	$135.00
2. Cost of insurance	136.50
3. First year's depreciation	402.00
4. Maintenance:	
a. Two new tires	51.36
b. Brake reline	23.40
c. Major tune-up	15.05
d. Renew front fender	49.55
e. Renew rear bumper	39.70
f. Adjust automatic transmission, change lubricant	7.90
First year cost of operation per mile	8.6c

HUDSON WASP TEST TABLE

PERFORMANCE

CLAYTON CHASSIS DYNAMOMETER TEST
(All tests are made under full load conditions)

RPM	MPH	ROAD HP
1200	30	29.5
2000	53	50.0
3000	80 (maximum)	60.5

Per cent of advertised hp delivered to driving wheels—47.5%

ACCELERATION IN SECONDS
(Checked with fifth wheel and electric speedometer)

	LO-D3	D-3	D-4
Standing start ¼ mile	:20.9	:21.3	:21.5
0-30 mph (0-34, car speedometer reading)	:06.4	:06.1	—
0-60 mph (0-67, car speedometer reading)	:19.0	:19.8	—
10-60 mph in DRIVE	—	:20.2	:24.3
30-60 mph in DRIVE	—	:15.8	:23.5

TOP SPEED (MPH)
(Clocked speeds over surveyed ¼ mile)

Fastest one-way run	103.5
Average of four runs	97.8

FUEL CONSUMPTION IN MILES PER GALLON
(Checked with fuel flowmeter, fifth wheel, and electric speedometer)

	D-3	D-4
Steady 30 mph	17.3	22.7
Steady 45 mph	15.5	20.2
Steady 60 mph	14.0	16.6
Approximate average in traffic	—	16.2

BRAKE STOPPING DISTANCE
(Checked with electrically actuated detonator)

Stopping distance at:	
30 mph	50 ft. 1 in.
45 mph	116 ft. 8 in.
60 mph	224 ft. 8 in.

SPEEDOMETER CHECK
(Checked with fifth wheel and electric speedometer)

ACTUAL	INDICATED	ERROR (%)
30	34	13
45	51	13
60	67	12

Odometer correction factor for 100 miles 1.0

GENERAL SPECIFICATIONS

ENGINE

Type	L-head
Bore and stroke	3$\frac{9}{16}$×4⅜ in.
Stroke/bore ratio	1.23:1
Compression ratio	7.2:1 6.7:1 optional
Displacement	262 cu. in.
Advertised bhp	127 @ 4000 rpm
Bhp. per cu. in.	0.484
Maximum torque	200 lbs. ft. @ 1600 rpm
Maximum bmep	115 psi

DRIVE SYSTEM

Transmission—Standard: Silent synchronized mesh, blocker type with helical gears. Optional: Hydra-Matic fully automatic drive; overdrive with standard transmission.

Rear axle: Semi-floating with alloy hypoid gears and alloy axle shafts.

DIMENSIONS

Wheelbase	119 in.
Tread	Front—58½, Rear—55½
Wheelbase/tread ratio	2:1
Overall width	77$\frac{3}{16}$ in.
Overall length	202$\frac{15}{32}$ in.
Overall height	60⅜ in.
Turning radius	19.65 ft.
Turns, lock to lock	5¾
Weight (test car)	3800 lb.
Weight/bhp ratio	30:1
Weight/road hp ratio	63:1
Weight distribution (front to rear)	44-56%

INTERIOR SAFETY CHECK CHART
(Same as Hudson Hornet)

OPERATING COST PER MILE ANALYSIS

1. Cost of gasoline	$126.00
2. Cost of insurance	143.50
3. First year's depreciation	365.00
4. Maintenance:	
a. Two new tires	77.48
b. Brake reline	26.27
c. Major tune-up	12.95
d. Renew front fender	31.85
e. Renew rear bumper	22.10
f. Adjust automatic transmission, change lubricant	6.30
First year cost of operation per mile	8.1c

was very evident over all types of roads from smooth cement to asphalt to brick.

Economical, But Not Miserly

On the recent Mobilgas Economy Run, the Hornet and Wasp both got averages of over 20 miles per gallon over a course that was more rugged than any previous Run. The best averages that MOTOR TREND Research was able to obtain were at 30 mph and were slightly better than the Run averages. The Wasp got two more mpg, while the Hornet got about ½ more mpg (see Test Table). This definitely indicates that both Hudsons are capable of good fuel economy. The surprising thing about both cars is that despite the fact that they have high compression ratios (7.2:1), they performed on Mobilgas Regular, not requiring premium gas. The only time we got any trace of ping was when the outside air temperature and the engine water temperature both became fairly high.

Wasp Hollywood—$2789.68 fob Detroit

Pacemaker Tudor—$2248.42 fob Detroit

Hornet Four-door—$2749.18 fob Detroit

Both the Hudsons tested were driven over long distances. The Hornet was taken on Economy Run trip; the Wasp went south through desert and over mountains

Plush Interior

HORNET: There are many things about both cars' interiors that we liked, some that we thought could be improved. The Hornet, in particular, has a plush-looking interior, appointments are all well-made, vision is good (but not under all conditions), the seats are exceedingly comfortable for long hauls, legroom and headroom are both better than average, but the over-chromed dash was found to be annoying.

The positioning of the steering wheel to the seat makes for good control and comfort, while the expected reduction in vision (because of the car's lowness, its width and the large expanse of the dash panel) was exceedingly easy to get used to. The only time that this is noticeable is when you're driving in traffic or attempting to park. Since you are not able to see the right front fender, you have to *feel* your way through traffic; you don't have the assurance of *knowing* where you are. And without raising up in the seat you cannot see the trunk lid (to judge distances when backing up). The left windshield post, where it meets the top (necessary because of the type of structure) restricts vision to some extent, particularly for mountain driving.

All controls are within comfortable reach, but it was difficult getting used to the window crank and the door handle—these are reversed in position from most cars. We certainly approve of the feature in the doors which do not allow them to be opened against the safety locks.

The dashboard is laid out quite simply and uses only red warning lights for the oil pressure and ammeter (to indicate low oil pressure and battery discharge). It's hard to tell how fast you're going because of the imitation "mother of pearl" behind the speedometer numbers. We understand that this is to be changed on later production models. Although a formerly bad spot of reflection (the top of the dash) has been taken care of by use of Dura-Fab (plastic leather), reflections are still bothersome because of the extreme use of chrome. Points of annoyance: the chrome facing around the instruments, the chrome glove compartment door, the chrome outside windshield trim, the windshield wiper and, occasionally, the chrome door ledge.

Because of the basic design of the car and the extreme angle of the rear window, cars become distorted in the rear view mirror, varying in shape from squat figures to exceedingly tall cars. Although this is a minor point it is annoying because it affects, to some extent, the ability to judge the speed of an overtaking vehicle.

Although we were never bothered on entering or leaving the car because of the step-down design, getting in and out is sometimes difficult for some women and short-legged men. Good features are the hand grips provided on the rear corners of the front seat to assist in getting out and the courtesy lights at each door.

MOTOR TREND Research certainly approves of the weather control unit, despite and because of the fact that the air entrance is in the cowl. The advantage of this is that when the fan is on, it will not suck in exhaust fumes of other cars while standing still in traffic, as is the case with some cars that have the intake openings in the front fenders. A very desirable feature from a standpoint of air circulation is that with the front window down and with the wind-wing open, there is no draft in the back seat.

WASP: Everything mentioned above in regard to the Hornet applies to the Wasp, with the exception of the following specific items: legroom, dash and rear view mirror.

As with the Hornet, the Wasp has plenty of headroom but legroom was found a little wanting in the front seat. The dash is set up just as it is in the Hornet but is more legible since imitation "mother of pearl" is not used for the speedometer; however, the speedometer needle would be more visible if it were painted a different color from the background. For night driving a rheostat dimmer for the instrument lights would be desirable, such as is used on the Hornet.

Our test Wasp was provided with a tinted windshield so that the glare that

Continued on page **30**

Hudson Road Test

The Hornet's engine has a horsepower per cubic inch rating of .47 while the Wasp is even higher with .48. Ratings indicate high efficiency powerplants

was noticed on the Hornet was largely eliminated, but some was still evident from the top of the dash. Even without the tinted windshield, there was one less glare point, since the door ledges are not chromed.

The stock rear view mirror provided with the car is not really wide enough to take advantage of the wide rear window. It was necessary to move around from side to side to take in the full road behind you.

It is a little awkward to crank the window up and down because your elbow hits the arm rest. On the other hand, the arm rest on the driver's side is useless as a resting place for the elbow if you sit directly behind the wheel. Placed out of the way in the door recess, along with the window crank and door handle, it provides for a few more inches of seating space, however.

Powerful Engine

Despite the swing of the industry toward overhead valve V-8 engine designs, Hudson is sticking to its L-head, in-line engines and is continuing to produce a powerful, efficient unit. The Hornet's .47 horsepower per cubic inch (a good indication of its efficiency) is one of the highest in the industry. Although the Wasp engine, a six-cylinder L-head, produces 18 less horsepower than the Hornet, its horsepower per cubic inch rating is even higher than that of the efficient Hornet, being .48. This engine is not the same block as that of the Hornet sleeved down, but it is an entirely new block. It is the same engine, however, that is used in the Commodore six.

The engines of both the Hornet and the Wasp, and of all Hudsons, have chrome alloy cylinder blocks, said by Hudson to be the toughest in the industry. The piston rings are pinned in position to prevent their rotation with the idea that this will prevent irregular wear of the piston rings and/or of the cylinder walls.

Each component of each engine is matched and statically balanced (by weight) before assembly. Then each engine is electronically balanced (balanced while running) as a unit to make it as smooth and quiet running as possible. For some reason, the engines of both test cars were not as quiet as might be expected. On starting, both were fairly noisy at first, then they would quiet down as they warmed up. Balancing certainly adds to smoother performance. The only indication of roughness was when we initially stomped on the throttle (at a standstill). The engine, at this point, had a terrific throb and vibration, which may have been due to the high torque and the type of engine mounts that are used.

Both engines are well laid out for easy servicing. The generator, distributor, starter and fuel pump are all located below the top of the block. Sufficient room, however, is provided around them to allow for easy servicing. The adjustment of the tappets would probably be a little difficult because of the location of the exhaust manifold, but even so it is not necessary to remove any fender panels to get at the valve cover.

The Hornet H-145 engine is the largest engine that Hudson builds, having a displacement of 308 cu. in. It develops 145 hp with a 7.2:1 compression ratio.

Brakes Have Reserve System

Brakes on the Hudson seem adequate for all conditions. On the thousands of miles we piled up on the cars, just once did the brakes give any indication of fading, and then only under the most severe usage in the mountains.

A very desirable safety feature of the brakes has been continued in 1952 models. It consists of a reserve mechanical system operating off the brake pedal. If the hydraulic system should fail, the mechanical system takes over by a harder pressure on the brake pedal.

Summing Up

Motor Trend Research had an enjoyable time testing both the Hornet and Wasp. Prior to obtaining the performance figures on the Hornet, the car was driven for 2500 miles over the Mobilgas Economy Run course (see July MT). Traveling with the Run cars, letting them all pass (for photographic purposes) then catching up and passing them, over the rugged terrain that the course followed took a car with lots of power and good handling characteristics. The Hornet definitely provided us with that. It is one of the most responsive cars that we have tested. As the public becomes more acquainted with the Hornet, Hudson should have difficulty building enough of them.

At its lower price, the Wasp is naturally not the car that the Hornet is, but dollar for dollar it is an equally good buy. When both the cars were driven at high speeds, through the snake-like curves and steep grades of the San Jacinto mountains, the Wasp proved its mettle. Under these conditions, it handled even better than the Hornet. Where the Wasp is better in this respect, the Hornet is better in acceleration, slightly better in top speed, and in the comfort department. Although it would be hard to choose between the two cars, it isn't hard to determine that either one of them is a good buy on their respective performance merits (see Test Table.)

—Walt Woron

HUDSON ACCESSORIES

Backup light	$ 8.95
Curb signals	1.75
Directional signals	22.75
Electric clock	26.95
Fog lights	14.65
Glare-proof mirror	23.50
Hydraulic jack	12.45
Karvisor	21.95
License plate frames	8.35
Radio & antenna	88.50
Spotlight & mirror	23.50
Underseat heater	19.50
Weather control	67.50

HUDSON'S JET PROMISES TOP PERFORMANCE

There is a completely new car in the low-priced field. It features conservative styling, but has a performance potential which promises to make other production models sit up and take notice

ONLY completely new car among 1953 U. S. production models is Hudson's Jet, designed for the low-priced field. Not all the facts about the car were immediately available, but last week a Jet was provided for detailed study so it became possible to assemble some interesting data about the new car.

Conservatively styled, the Jets feature many principles of design and construction which have been fully tested on the Hudson Hornet and Wasp. However, incorporating the features into the compactly built Jet involved a considerable investment by Hudson in tooling and the car is genuinely new, not a rehash of components from other models.

Two series make up the line—the Jet ($1,685) and the Super Jet ($1,775). Both are identical except for minor differences in trim and accessories. The "step-down" design, characteristic of other Hudsons, has been retained and the body is of boxed steel sections, with floor recessed within frame members. Road holding ability, traditional with Hudsons, is one of the performance features stressed. Total dry weight is 2800 pounds.

FRONT suspension consists of independent coil springing, with direct-acting shocks vertically mounted inside coils at center of spring action. At the rear, semi-elliptical leaf-type springs are positioned at angle for improved stability and rear shocks are diagonally mounted.

Engine is rated at 104 hp to 114, depending upon optional equipment which includes aluminum head of 8:1 compresion and Twin-H Power carburetion. Normal compresion ratio with cast-iron head is 7.5:1. Block is of high chrome alloy.

To date the Jet has not appeared in stock car competition. Evaluating its potential produces some formidable figures, however. At maximum hp and at factory listed weight, the Jet has a power-to-weight ratio of 24.5 to 1, which rates it extremely high among U. S. mass-production autos.

SEMIFLOATING rear axle with hypoid gears has a standard gear ratio with manual transmission of 4.10:1. A special ratio of 4.27:1 is optional for mountainous areas and Hudson offers a 3.31 ratio for economy. Overdrive has 4.27 to 1 ratio as standard and 4.10:1 optional. With Hydra-Matic drive, 3.54:1 is standard and 3.31:1 optional. Brakes are hydraulic, steering gear is needle-mounted worm and roller type and tire sizes are 5.90 x 15 on Jet, 6.40 x 15 on Super Jet.

Balance of the equipment on the Jet consists of familiar Hudson items, such as weather-control heating and ventilating, flashing oil pressure and generator lights. Tinted safety glass is optional at extra cost.

Photo at top of page shows frontal view of Super Jet, while lower picture is of rear end of standard Jet. Gas intake pipe is centrally located just behind bumper.

Dash of Super Jet has familiar Hudson items. Hydra-Matic Drive is optional at extra cost. Seat rises as it is moved forward.

Engine of Jet is particularly important, gives car maximum of 114 hp, high power-to-weight ratio.

No. 1504: HUDSON SUPER JET SALOON

A four-light arrangement with clean, simple lines forms the basis of the Hudson body style. To emphasize the impression of length, a lower "crease" supplements the falling wing line.

The Autocar ROAD TESTS

THAT it is small by American standards is the first impression created by the Hudson Jet, yet it is not a transatlantic attempt to produce a small car, but rather one of a medium size with a very good performance, somewhat on the lines of post-war six-seater British cars, and smaller than the Hornet and Wasp models which, together with the Jet, make up the Hudson range. The Jet is, in general conception, of similar overall proportions to some European models, yet, as in all the Hudson range, the power unit is a six-in-line side valve as distinct from the general trend in America, where the overhead valve V layout is becoming very popular. The 3,310 c.c. engine is available in two forms: the Jet, which has a single carburettor manifold and develops 104 b.h.p., or the Super Jet, which, with the aid of two downdraught Carter carburettors and high compression ratio cylinder head, develops an extra 10 b.h.p. The latter model has recently been tested on the Continent by *The Autocar*.

The car provided for test was fitted with optional equipment in the form of a dual-range Hydra-Matic transmission, similar to that fitted by one British manufacturer on some export models. With the two-carburettor engine the car has a very lively performance as regards both acceleration and maximum speed. Having a mean maximum speed comfortably above the 90 m.p.h. mark, the Super Jet can be cruised all day at 75-80 m.p.h. without any sign of fuss, and even at that speed there is a very useful reserve of power to accelerate the car to its top speed of 92 m.p.h. mean, where it feels both comfortable and safe.

The engine is smooth and quiet, and gives the impression that it is rugged and tough. It has the minimum of moving parts and consequently requires very little maintenance. On ordinary grade fuel some pinking was noticed if the engine was pulling hard, but this was completely eliminated when running on first-grade fuel, which was used for the performance tests. The transmission is the latest version of the well-known Hydra-Matic, and, in place of the single position on the selector lever of the original system for drive range use, two positions are now provided, Dr 4 and Dr 3.

Briefly, the mechanism consists of an hydraulic coupling, and a planetary gear box providing four forward speeds. Apart from the drive range positions, the selector lever also covers a low range which provides drive on first and second gears only. If the car is accelerated hard (with the throttle fully opened) the transmission then stays in first gear. On part throttle a change up to second gear is performed automatically at a speed of 40 m.p.h., while the mechanism changes down to first gear if the speed falls below 30 m.p.h. With the lever in the Dr 3 position the car starts in first gear, and if the throttle is wide open changes occur automatically at speeds of 18, 28 and 63 m.p.h. In Dr 4 position the change up into second and third gears occurs at the same speeds as in Dr 3, but the change into top gear takes place at 50 m.p.h. If the throttle is closed, and the car is allowed to decelerate, the transmission automatically changes down at speeds of 12, 6 and 4 m.p.h.

Between these limits of speed, gear changing is automatic, depending on loading and throttle opening. The advantage of the Dr 3 position is that third gear can be selected at a speed greater than is possible in Dr 4, thereby increasing the

The doors are thin and open wide to permit easy access to both compartments. Combined pulls and arm rests are built into all the doors, and there is a rope "rug rail" below the ash tray on the back of the front seat. Because of the low build of the car, there is a propeller-shaft tunnel in the floor.

The frontal treatment is simple and functional. A low wide air intake extends across the full length of the body and encloses the indicator lamps. The motif and small grille on the top of the bonnet are a style feature, and not an air intake to the engine. The large rear window space can be appreciated from this view. The rear treatment is nicely balanced. The roof line and large curved rear window blend well with the luggage locker. Domed rear lights are built in to the finned ends of the wings. Push-button door locks are used, and both front doors can be locked with a key.

ROAD TEST

acceleration when necessary without adversely affecting the fuel consumption.

With the aid of a fluid drive, the take-up from standstill is very smooth, and the change from gear to gear can be performed without snatch. The transmission is quiet in operation and provides a very satisfactory means of two-pedal control. It is possible to start the engine by turning the ignition switch in a clockwise direction past the normal position. This can be done, however, only when the selector lever is in neutral, and serves as a safety measure to prevent the car from moving as soon as the engine fires.

Experience of various 1953 American cars has created the impression that the general riding and suspension qualities of vehicles produced in that country have considerably improved. This impression was confirmed by the Hudson. Over a wide variety of road surfaces, including smooth concrete, rough "colonial" sections and Belgian *pavé*, the suspension produces a level and very comfortable ride in both front and rear seats, free from shocks and pitching. The springing is sufficiently soft to cope with rough roads, yet it is well controlled and the dampers do not show signs of fade after many miles on rough stone setts.

In keeping with the general roadholding, there is very little roll on corners. The car is directionally stable, having a useful degree of understeer. In view of its 4½ turns from lock to lock, the steering is not quite so light as would have been expected, but this is compensated for by the general precision with which the car can be controlled. Road shocks are not transmitted to the driver's hands, but it is possible to feel what is going on at the front wheels, a useful feature when driving under adverse conditions.

The hydraulically operated brakes are very powerful, and under test conditions on dry concrete they recorded a very good maximum efficiency despite relatively light pedal application. No fade or loss of balance was experienced either on the road or during the protracted test procedure which requires more frequent and forceful brake application than normal operation.

The general noise level is very low, but there is some noise particularly when driving on stone setts. As regards wind noise the Super Jet is very quiet, and even when driving fast with the quarter lights or side windows open the noise is not excessive.

A well-known Hudson feature of placing some of the stress-carrying members outside the rear wheels helps to reduce the body height in general and the floor level in particular; as a result, the seats in the Jet saloon are quite low, but still a reasonable height from the floor. On the driver's side, the floor seems quite empty, containing only two pedals and a small dip switch which is designed to be operated by the driver's heel. The pedals and steering wheel are well positioned in relation to the driving seat, which is of a comfortable shape and well sprung.

One of the most useful assets of the low build, from the driver's point of view, is the ability to see both wings. The general all-round visibility is also good, with windscreen pillars that produce a minimum of blind spots, and a very large wrap-round rear window, which in conjunc-

The six-cylinder engine is completely surrounded by its auxiliaries. Twin air cleaners are used for the two carburettors of the Super Jet. The battery is placed to the left of the engine and can just be seen behind the radiator filler cap.

The large luggage locker also houses the spare wheel and tools. The floor is covered with protective carpet to prevent damage to luggage. A guttering on the inside prevents water running into the compartment when the lid is opened.

tion with a well-placed rear view mirror provides the driver with both long and short range visibility to the rear.

The instrument layout is both simple and sensible, with the minimum number of instruments to distract a driver's attention, while the important items such as loss of oil pressure or dynamo charge are brought to the notice of the driver by means of red lights. The instruments include water temperature and fuel gauges as well as the speedometer, and are grouped in front of the steering wheel. A cowl is fitted to prevent reflections in the windscreen at night. This screening is very effective in front of the driver, although some reflection does occur on the opposite side of the facia, caused by decorative chromium on the facia locker lid. Suction-operated windscreen wipers are used, and, although they cover a useful area of the screen, they tend to stall when the throttle is wide open.

For a medium-sized car the luggage locker is very large. The spare wheel is mounted vertically inside the locker, on the right-hand side, but even so there is still ample space for luggage. The fuel tank is placed below the locker floor and filled via a flap in the rear body panel. The tank can be replenished quickly without risk of blowing back, and with its 12-gallon capacity the car has a useful range between fuel stops. Double dip head lamps on the car tested provide a good spread of light, although a longer beam would be useful for fast night driving. The horns, operated by a "C" ring mounted on the steering column, are effective and have a penetrating note. Thirty-three chassis lubricating points require attention at intervals of 1,000 miles.

The Hudson Super Jet is in many ways a car with a European character. It is trim and compact, has a good performance, and is well finished.

HUDSON SUPER JET SALOON

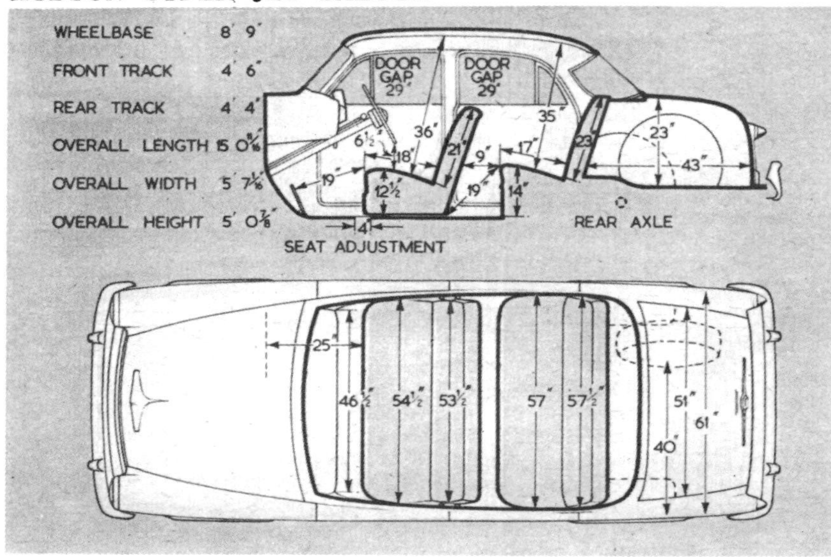

Measurements in these ¼in to 1ft scale body diagrams are taken with the driving seat in the central position of fore and aft adjustment and with the seat cushions uncompressed.

DATA

PRICE (basic), with saloon body, 149,900F (Belgian) = £1,070 at 139.6F = £1.
Extras: Radio 5,500FB = £39 (approx.).
Heater 4,500FB = £32 (approx.)

ENGINE: Capacity: 3,310 c.c. (216 cu in)
Number of cylinders: 6.
Bore and Stroke: 76.2 × 120.6 mm (3 × 4¾in).
Valve gear: side.
Compression ratio: 8 to 1
B.H.P.: 114 at 4,000 r.p.m. (B.H.P. per ton laden 78.6).
Torque: 168 lb ft at 1,800 r.p.m.
M.P.H. per 1,000 r.p.m. on top gear, 20.5.

WEIGHT (with 5 gals fuel), 25½ cwt (2,858lb).
Weight distribution (per cent) 52.7 F; 47.3 R.
Laden as tested: 29 cwt (3,258lb).
Lb per c.c. (laden): 0.985.

BRAKES: Type: F, Two-leading shoe. R, Leading and trailing.
Method of operation: F, Hydraulic. R, Hydraulic.
Drum dimensions: F, 9 in diameter, 2in wide. R, 9in diameter, 2 in wide.
Lining area: F, 66.1 sq in. R, 66.1 sq in. (91.3 sq in per ton laden).

TYRES: 6.40 — 15in.
Pressures (lb per sq in): 24 F; 22 R.

TANK CAPACITY: 12½ Imperial gallons
Oil sump, 9 pints.
Cooling system, 25 pints (plus 2 pints if heater is fitted).

TURNING CIRCLE: 33ft 5 in (L and R).
Steering wheel turns (lock to lock): 4½.

DIMENSIONS: Wheelbase 8ft 9in.
Track: F, 4ft 6in; R, 4ft 4in.
Length (overall): 15ft 0 11/16in.
Height: 5ft 0⅞in.
Width: 5ft 7 1/16in.
Ground clearance: 8in.
Frontal area: 22 sq ft (approx).

ELECTRICAL SYSTEM: 6-volt, 90 ampère-hour battery.
Head lights: Double dip, 35-35 watt.

SUSPENSION: Front, Independent; wishbones and coil spring; anti-roll bar.
Rear, Half-elliptic springs.

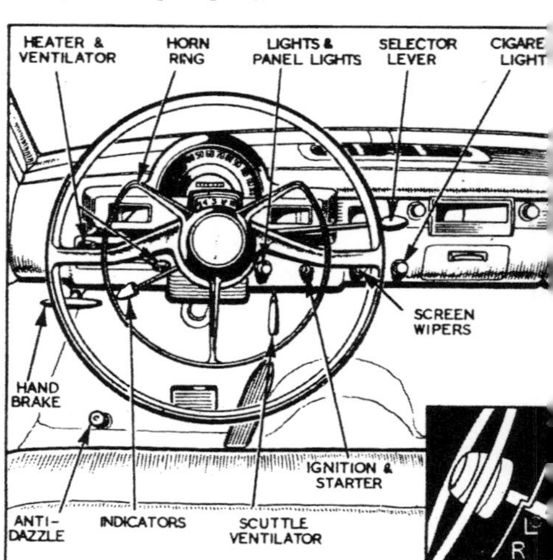

PERFORMANCE

ACCELERATION: from constant speeds. Speed, Gear Ratios* and time in sec.

M.P.H.	DR4 range	DR3 range	L range
10—30	4.7	4.6	4.2
20—40	7.1	6.6	5.1
30—50	7.6	7.2	—
40—60	9.1	7.7	—
50—70	12.7	10.1	—
60—80	14.6	—	—

*Gear ratios 3.54; 5.13; 9.32 and 12.52 to 1.
From rest through gears to:

M.P.H.	sec
30	4.5
50	10.7
60	15.1
70	21.8
80	32.2

Standing quarter mile, 20 sec.

SPEED ON GEARS

Gear		M.P.H. (max.)	K.P.H. (max.)
Top	(mean)	92	148.1
	(best)	93	149.7
3rd		62	100
2nd		28	45
1st		18	29

TRACTIVE RESISTANCE: 20 lb per ton at 10 M.P.H.

TRACTIVE EFFORT:

	Pull (lb per ton)	Equivalent Gradient
DR4	250	1 in 8.9
DR3	350	1 in 6.3
Low	553	1 in 3.9

BRAKES:

Efficiency	Pedal Pressure (lb)
96 per cent	94
51 per cent	50
30 per cent	32

FUEL CONSUMPTION:
18.9 m.p.g. overall for 120 miles (14.9 litres per 100 km).
Approximate normal range 17-20 m.p.g. (16.6-14.1 litres per 100 km).
Fuel, Belgian first grade.

WEATHER: Fine, dry surface; very slight wind.
Air temperature 68 degrees F.
Acceleration figures are the means of several runs in opposite directions.
Tractive effort and resistance obtained by Tapley meter.
Model described in *The Autocar* of June 19, 1953.

SPEEDOMETER CORRECTION: M.P.H.

Car speedometer	10	20	30	40	50	60	70	80	90	100	103
True speed	10	19	28	36	45	54	63	72	80	90	93

ROAD TEST: TWIN HORNET

Fifth wheel bobbing along behind, the Hudson digs in for an acceleration run.

By Barney Clark

WHEN you get out of the cities, out on the long straights and the winding mountain passes, you notice a change in the proportional ratio of makes of cars you meet. For one thing, you seem to see a lot higher percentage of Hudsons.

We don't have any actual statistics to support this, but it stands to reason. The traveling salesman with a big territory to cover, the executive who wants to get from Point A to Point B in a hurry, the long distance vacationer who rolls from coast to coast—they all want a big, rugged, comfortable car that will hold high cruising speeds *safely*. That, for many of them, means Hudson.

The reasons for this were demonstrated during ASR's recent road test of the 1953 Hudson Hornet—with Twin H power. There is nothing startling in the specifications of the new Hudsons; the cars are little changed over previous models and, indeed, they might almost be called old-fashioned. The Hornet engine is a flathead, in-line six with 145 advertised horsepower; the transmission on the test car was the well-tried Hydra-Matic; suspension is the conventional independent coil front and semi-elliptic leaf rear. Weight distribution is normal, with the front wheels carrying a higher percentage of the load.

Nothing here to indicate any particular virtue. But when you consider that the Hudson Hornet is the national stock car champion, a favorite mount of men who earn their livelihood by winning races, holder of every AAA record in stock competition, you conclude that somebody blended these conventional elements into a singularly well-balanced whole.

FIRST impression of the Hornet is the driving position, which rates A-1 in the American league. The wheel is big (18 inches) and set at a comfortable angle; the backrest comes firmly up behind the driver's back and shoulders to minimize sliding about on fast turns; seats themselves are high to give a natural position and reduce the fatigue that comes from cramped posture; throttle and brake take the foot at a comfortable angle. Instruments are sensibly grouped in front of the driver but, alas, are surrounded by lavish bands and

Hornet didn't "nose-dive" badly under severe braking, held close to a straight line even with wheels locked. Air scoop on hood is strictly for looks, has absolutely no function.

TEST TABLE: 1953 Hudson Twin H Hornet

Engine specifications	
Type	L-head
Bore and stroke	3-13/16 x 4-1/2 in.
Compression ratio	7.2 to 1
Displacement	308 cu. in.
Advertised b.h.p.	145 at 3800 r.p.m.

Drive Specifications

Transmission: Dual Range Hydra-Matic with four speeds forward (optional equipment); standard: manual three-speed synchromesh (overdrive optional). Rear axle: semi-floating hypoid.

Dimensions	
Wheelbase	124 in.
Tread	58½ in. front; 55½ in. rear
Length	208½ in.
Width	77¼ in.
Height	60-3/8 in.
Steering ratio	25.6 to 1
Turning radius (left)	20 ft. 5 in.
Weight	3973 lb.

PERFORMANCE
(Fifth wheel speedometer and stopwatch)

Standing Start (Drive Range only)	
0-30 m.p.h.	5.2 sec.
0-50	11.4
0-60	15.4
0-70	21.0
0-80	29.9

Constant Speed Start	
10 to 30 m.p.h.	3.8 sec.
30 to 50	6.8
50 to 70	11.1
Standing quarter-mile	20.1
Top speed	99.8 m.p.h.

Speedometer Comparison

Fifth wheel	Hudson speedometer
20 m.p.h.	21 m.p.h.
30	30
40	40
50	51
60	61
70	72

HUDSON HORNET

strips of chrome. Anybody lights a match in the Hudson at night, he can blind the pilot.

For years Hudson has claimed the widest seats in the industry (64 inches) and we aren't going to argue with them. The width is given emphasis by Hudson's trick of recessing arm rests, door handles and window-winders into the door thickness. You all know about the "step-down" design now, so we won't go into that, other than to remark that it imposes no real inconvenience getting in or out and does get the center of gravity down.

Hudson's bulbous body styling makes for big radiuses on door and window corners and for smaller window and windshield openings, so that a first-time passenger gets a faint touch of claustrophobia. But actually the driving vision is good (except for overhead traffic lights) and the only real complaint is about the back window. This, due to its deep curve, has a "lens" effect and distorts cars behind, making judgement of their exact distance difficult.

WHEN you first get the Hornet underway, the initial impression is one of size and solidity. There is no power steering and the ratio is rather high, so the wheel requires a good many turns on sharp corners. However, the effort is light (except while parking) and there is an unusually strong feel of solidity and *response* for an American car. Because of the swollen sides of the car and the inability to judge accurately exactly where you are in close-packed traffic, you are inclined to give your neighbors on the road too much room, just to be on the safe-side.

The brakes are firm and moderately light, with no unpleasant "wrapping" effect. Our earlier experience with Hudsons lead us to anticipate a rather harsh motor, in which the low-speed power impulses really made themselves felt, but the Hornet's engine was unexpectedly smooth and—almost—silky. The factory matches and staticaly balances each moving part before the big engines are assembled and then electronically balances the running engine, which probably accounts for the smoothness of what should be, theoretically, a rugged-running powerplant.

Our first informal acceleration tests were not particularly impressive, even with the Twin H installation (which consists of two single-throat carburetors for easier breathing, plus a slightly hotter cam). Such an installation was not expected to contribute much to low-speed torque, but even at higher r.p.m. the take-off was not astonishing. Part of this was undoubtedly due to the fact that the test car was the first 1953 Twin H Hornet in the area and had to be tested without full tuning. But, as the tabulated acceleration figures indicate, the Hornet is not the most potent of the high-performance American cars (nor is it the fastest) and the secret of its racing success lies in other factors.

What the major factor is we discovered as soon as the Hornet was pointed down the winding back-country road ASR uses for handling tests. This 25-mile strip is narrow, fairly high-crowned blacktop, with curve following curve in almost endless succession, interspersed with short grades, long, winding descents and short straightaways.

On test day this stretch was greasy with intermittent rains—a real nightmare with a car that won't handle. But the Hornet clung to this roller-coaster with a tenacity many a European car might envy. Pushed really hard on the wet surface—hard enough to bounce the rear-seat passengers from side to side—it only lost its grip once or twice and then didn't "skate." It merely slid a few feet, grabbed traction again and took off. Tire adhesion was remarkable even under sharp, hard braking and the steering wheel gave real correction in slides.

The major fault noticed on this type of road was when a really sharp corner cut the speed way down. Then the third gear in the Hydra-Matic transmission didn't put out enough torque and it dumped down into second. Second came in with a boom and a surge, the rear wheels spun and the driver longed for a higher ratio on this gear—and the ability to select it when, as, and if desired.

HIGHSPEED runs on narrow two-lane straightaways illustrated another virtue of the Hornet—it wandered far less than most American cars and was far easier to keep on its side of the pike. The sight of an approaching vehicle didn't cause an involuntary contraction of the stomach muscles and a relaxation of the throttle foot. If the slight initial mushiness in the steering could be eliminated, so that the response was as immediate in straight-line steering as it is in actual turns, the Hudson wouldn't need to fear comparison with some of the best of European cars.

The brakes, used hard and often on the 25-mile test road, were top grade for about 15 miles. Then fade set in, despite the coolness of the day, and the pace had to be moderated to cope with this. Recuperation from fade didn't seem particularly rapid.

In the actual acceleration tests, the Hydra-Matic produced the usual irritation, shifting where and when it wanted to. Since most drivers are probably going to order this transmission, however, it seemed the most likely set-up for the test. The dual-range element, which allows the driver to select third whenever *he* wants it and locks out fourth, inspired thoughts about how nice it would be if all four gears could be manually chosen (a la Rolls-Royce) and the automatic feature used only when the driver wanted to be lazy.

Experiments using starts in Low Range, with manual shifts to High Range, failed to show much actual gain over normal shiftless operation, so the tabulated figures are for straight High Range starts. There isn't much question but that these figures could be improved with the three-speed manual transmission, particularly the 0-30 m.p.h. bracket. In fact, using stopwatch-against-speedometer with an earlier model Hudson, a figure of 3.2 seconds had been recorded, as contrasted against 5.2 with the fifth-wheel speedometer (that Hudson clutch really bangs in and takes hold on dynamite starts and speed-shifts).

IT IS apparent that, with Twin H power, the Hudson is in the growing class of American cars that can top the century mark. The test car didn't, due to (a) lack of tuning, (b) moist road surface, and (c) insufficient run-in distance. But 99.8 m.p.h. was close enough to indicate it can be done—and has been done under more favorable conditions.

Right here might be the place to point out that the Hudson speedometer is one of the most accurate in the trade. And that's mighty refreshing after years

Continued on page 43

Twin H chromed air cleaners give engine big look.

Dash (above) has more chrome than it needs; trunk (below) is ample. Spare tire is mounted vertically.

ROAD and TRACK ROAD TEST No. A-2-53

Hudson Super Jet

SPECIFICATIONS

Wheelbase	105 in.
Tread—front	56.5
—rear	54.0
Tire size	6.40 x 15
Curb weight	2980 lbs.
weight on rear	45.3%
Test weight	3280 lbs.
Turning circle	33.4 ft.
Turns lock to lock	4.2
Engine	6 cyl. in line
Valve system	side
Bore & stroke	3.00 x 4.75 in.
Displacement	202 cu. in. (3311 cc.)
Compression ratio	7.5
Horsepower at 4000 rpm	104
Torque at 1600 rpm	158 ft. lbs.
Transmission—	3 speeds with overdrive
Gear ratios—overdrive	2.99
—high	4.27
—2nd od	4.87
—2nd	6.95
—1st od	7.79
—1st	11.12
Mph per 1000 rpm (in od)	26.3
Mph at 2500 fpm piston speed (in od)	83.3
R & T performance factor (high gear)	52.2
Seating capacity	6
Price at factory (with od)	$2120

PERFORMANCE

Top speed (avg.)	88.49 mph
Fastest 1-way	91.09 mph
Max. speed in gears	
—high	84.9 mph
—2nd od	83 mph
—2nd	63 mph
—1st od	54 mph
—1st	37 mph
Shift points from	
3rd	84 mph
2nd	54 mph
1st	36 mph

TAPLEY READINGS

Pulling Power	Gear	MPH
595 lbs/ton	1st	26
450 lbs/ton	2nd	30
305 lbs/ton	2nd od	36
292 lbs/ton	high	40
190 lbs/ton	od	43

(COASTING)

40 lbs/ton at	10 mph
65 lbs/ton at	30 mph
110 lbs/ton at	60 mph

SPEEDO ERROR

Indicated	Actual
10 mph	11.3
20 mph	20.0
30 mph	28.4
40 mph	37.2
50 mph	45.2
60 mph	54.3
70 mph	62.5
80 mph	71.5
90 mph	81.5
100 mph	91.0

ACCELERATION

0-30 mph	3.46 secs.
0-40 mph	6.37 secs.
0-50 mph	9.65 secs.
0-60 mph	15.14 secs.
0-70 mph	19.98 secs.
0-80 mph	29.41 secs.
Standing ¼ mile (avg.)	20.10 secs.
Best standing ¼ mile	19.81 secs.
Mileage	19.5/23.2 mpg.

Photographs by Chesebrough

Our approach to the new Hudson Jet was one of mild interest—perhaps even condescension. When we completed the road test our general opinion was completely reversed. No fair minded driver can give this car a trial without coming to a definite set of favorable conclusions for here is a scrappy little car that is going to give both the big 3 (Chevrolet, Ford and Plymouth) and the little 3 (Henry J, Rambler and Willys) some real competition.

While the Jet is more compact than the big 3 cars it concedes nothing in roominess, with 58 inch seats front and rear. It weighs around 300 lbs. less than the big 3 average yet loses absolutely nothing in the way of riding comfort and gains appreciably in the performance, economy and handling departments because of this weight saving.

PERFORMANCE . . .

While the Jet name may be optimistic (compare performance with the Rover "Jet" road test, this issue), our hard-boiled Tech. Ed has this to say: "After driving the Hudson 202, I take back what I said about this car (March *Misc. Ramblings*). It's a Honey Bee alright—a 'honey' and a bee with a real sting."

The performance data and acceleration curve speak for themselves, and frankly we didn't believe the figures—at first. We checked them again and again, and then began comparing Tapley readings. It outpulls any of the 6 comparable cars in each gear. Then we tried side by side acceleration from a standstill. The Jet will leave any of the 6 cars at the signal lights.

During the timed top speed runs (we made 6 runs) the speedometer floated at around an indicated 100 when 91.09 was recorded. Subsequently we found that we could just keep up with one of the fastest of the big 3 cars in 2nd overdrive, and pass him in direct drive. Actually at 91 mph in overdrive the engine is turning only 3460 rpm, or 500 below the peak horsepower speed. At the 85 mph maximum in direct drive the engine turns 4610 rpm or 600 rpm beyond the optimum revolutions.

These figures show that a compromise of gear ratios has been achieved, at a slight sacrifice in all-out top speed. Direct drive gives very good acceleration, overdrive gives very creditable economy. A ratio between these two gears would give better top speed with a drop in both acceleration and economy. We did find that acceleration in the 2.99 overdrive ratio was very poor, a fault accentuated by a kickdown switch that took a lot of pressure to activate. One of the manual overdrive switches now available would completely eliminate this objection.

As a matter of minor importance it has

The Hudson Super Jet alongside a well known competitive car. Though smaller overall, the Jet has ample room for 6 passengers, rides very well and gives light handling. Wheelbase is 10 inches less, yet the overall length is shorter by 17 inches.

been estimated that the optional equipment of 8.0 aluminum head, twin carburetors, hydramatic transmission and 3.54 axle ratio will add close to 5 mph to the top speed.

COMFORT AND HANDLING . . .

Lacking electronic equipment to rate the Jet's riding qualities, we can only say that it is amazing in view of the 105 inch wheelbase and as far as we can tell, is the equal of any other car in the low price field.

The steering is very light, and surprisingly quick for an American car. Even at high speed and with gusty side winds the car felt safe and it was not tiring to drive. Remarkable too was the fact that there was no over or understeer, nor was there any tendency for the rear end to break loose on fast corners. Our only complaint on the handling qualities was a rather high degree of initial lean on severe corners. Stiffer shocks would help here as well as reduce tossing on high speed road dips.

"With its low center of gravity, the new Hudson Jet is steady on the road at all speeds. Like all Hudsons, it hugs the curves as no car ever has before."—From Hudson Catalog.

ENGINE . . .

The Hudson company has never been known for conventionality and the entirely new 202 engine represents a school of design well opposed to the trend of the past few years toward oversquare, overhead valve engines. Since the tooling costs alone, for a new engine usually exceed one million dollars, the long stroke, side valve design must have some merit. The arguments in favor of this design go something like this:

1. The small bore gives very light reciprocating weight so that rod bearing loads are little different from those of a large bore short stroke engine.
2. Main bearing loads can be controlled by counterweights.
3. The small bore gives high thermal efficiency.
4. The small bore permits higher compression ratio.

(Continued on page 75)

The L-head Jet engine is very accessible for tune up and maintenance, especially when compared to ohv V-8 engines. Cylinder bore of 3 inches gives short unit.

Storming down the highway in 2nd overdrive, with needle indicating a very optomistic 90 mph. See text for the details.

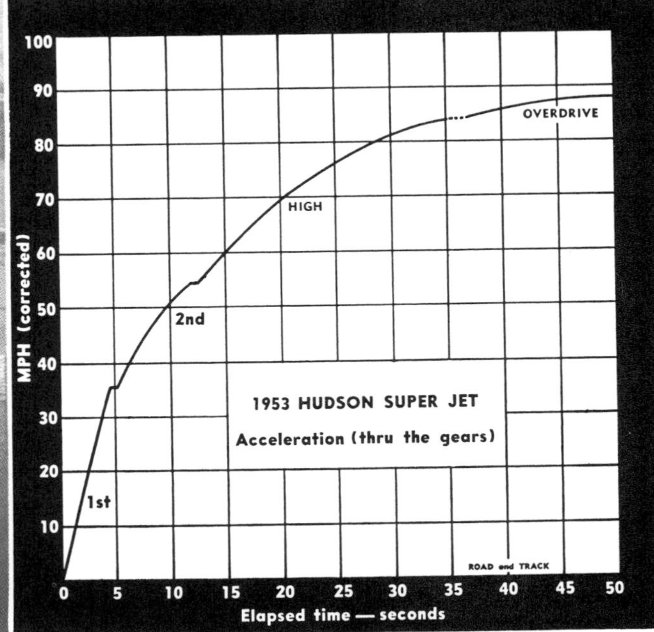

Good acceleration from a standstill is graphically illustrated. All the data is averaged, run on a corrected speedometer.

Acceleration that leaves most Detroit iron gaping at its bustle back, and economy to make a miser envious, mark Hudson's entry in the low-price field.

Due to inclement weather during the test, this Super Jet was borrowed from Old Dominion Motors of Alexandria, Va. for exterior picture purposes.

ROAD TEST

Super Jet

power plus—30 mpg

SPEED AGE RESEARCH REPORT

PHOTOS BY DON O'REILLY AND DICK ADAMS

HUDSON'S Super Jet won't win any beauty contests but its more buxom sisters needn't cheer—in the talent department the 'ugly duckling' seems destined to wind up with the automotive world's Miss America title for 1953.

This slender offspring of the famed Hornet is an automobile of big car performance and small car economy which has planted its four wheels firmly in the $2,000-plus market.

Fitted with high compression head, dual carburetion (Twin-H-Power) and Dual Range Hydra-Matic, the Super Jet's acceleration packs enough wallop to leave most Detroit iron gaping at its bustle back.

Couple this with comfort, roadability, and better than 30 miles per gallon at certain speeds and the competition has cause to fret.

The powerplant of this bundle of dynamite is nothing new—rather it is a scaled-down version of Hudson's tried and proven L-head Six. Ignoring the trend to oversquare engines, the company used a 3-inch bore and 4.75-inch stroke for a displacement of 202 cubic inches—less than any of its Big 3 rivals.

With 8 to 1 aluminum head, and Twin-H-Power, this unit puts out 114 HP. Standard trim calls for a 7.5 to 1 head which delivers 104 HP. Another option of dual carburetion on the standard engine produces 106 HP.

A portion of the credit for the terrific performance can be attributed to a road weight of approximately 3,000 pounds, achieved by following Hudson's Monobilt body and frame design which eliminates parts of the heavy and conventional chassis without sacrificing rigidity. The 'step-down' in the Jet, however, is not as pronounced as in the Hornet.

Gone, too, are the wide shoulders, so to speak, of the Hornet and Wasp. Interior specifications of the Jets reveal that the front seat is only 54 inches wide at shoulder height, compared to better than 60 in the parent models. This scaling down applies to all the dimensions; wheelbase is 105 inches, tread 54 inches and overall width a bit more than 67 inches.

Interior finish follows closely the superior workmanship usually associated with Hudson products with upholstery which is more than serviceable and Durafab trim. The seats are of chair height and soft, but provide sufficient support that long-distance driving is not tiring. Legroom, however, is scanty in the rear.

The instrument panel, unlike other Hudsons, is free of excessive chrome. The speedometer, mounted high on the dashboard, is the centerpiece and is flanked by temperature and fuel gauges. Lack of oil pressure and failure of the generator is indicated by flashing red lights. Radio and ashtray are in the center of the panel and within reach of the driver but the glove compartment is still at the extreme right.

Body-wise the Super Jet and its lower-priced running mate, the Jet, has few esthetic points. The stylists could have drawn this one with a straightedge for there are no false or flattering curves. Except for the fraudulent airscoop on the hood, the design is as functional as a box and covers the interior without pretenses.

Hudson Motor Car Company delivered the test car, a Super Jet with all optional engine equipment and Dual Range Hydra-Matic, to Rodocker Motors, Inc. at Indianapolis, Ind. where it was picked up by members of the SPEED AGE staff. Then the fun began and continued for more than 2,000 miles until the automobile was returned to Indianapolis.

That 2,000 miles included every conceivable type of roadbed from the dual highways of Ohio to twisting, winding Route 40 over the Alleghenies in southern Pennsylvania and Maryland. The first 600 miles, into Washington, D. C., showed a gas consumption equal to 22 miles per gallon with a driving time of slightly less than 12 hours. Eyebrows went up, pencils came out and the figures were rechecked then set aside to await the more accurate results of a mileage meter test.

The full enjoyment of driving this car, however, comes at the traffic lights when Dragstrip Harry belatedly realizes the ugly duckling he dismissed with a disdainful glance has more scat than his super-horsepower chariot.

Although a comfortable over-the-road automobile, the full measure of the Jets is appreciated in city traffic where the short wheelbase and fast steering takes much of the work out of driving. More slender than most of its competitors, it can go through narrower holes, is more maneuverable, and glides into those hard-to-get-into parking places.

The steering—four turns, lock to lock—combined with the flexibility of the engine and Hydra-Matic also make the Super Jet an excellent mountain goat. The hairpin turns of Route 40 through western Maryland were run deliberately at speeds 20 MPH above those posted without any indication that the Jet desired to wag its tail.

There is, as with all cars having coil spring front suspension, a reasonable amount of body roll in tight corners. However it was not difficult to keep the Jet within its own lane, even under those conditions. The short wheelbase does permit considerable bobbing of the front end on rough roads, particularly at low speeds. This smooths out as the speed increases and the ride improves. But wind wander is a problem and the Jet's comparatively high silhouette makes it ex-

Both front and rear doors open wide, making access easy for the passengers.

The sparing use of chrome trimming on the instrument panel reduces glare.

Air scoop on the hood is a falsie, just about the only non-functional part of the design. The lines, although boxy, are strikingly honest and without frills.

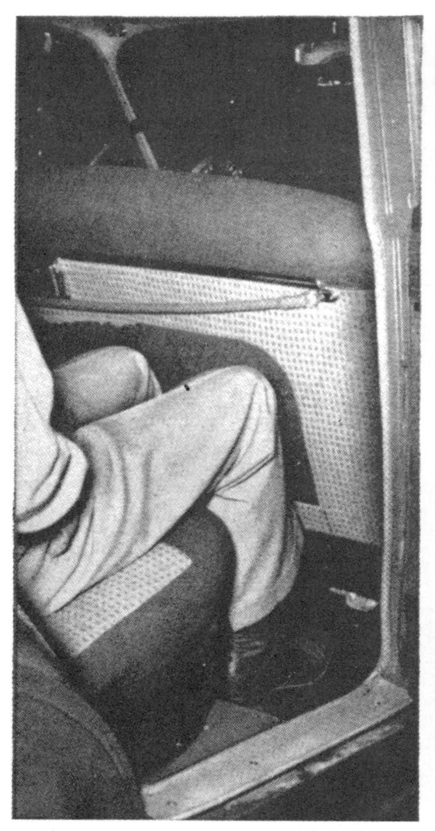

Don O'Reilly Photo

The stubby body design leaves little legroom for the rear seat passenger.

Don O'Reilly Photo

Dual air cleaners of Twin—H-Power must be removed to change spark plugs.

Don O'Reilly Photo

Belying its exterior appearance, trunk space is adequate. Covering the seat back lends a quality touch. Well braced trunk lid is counterbalanced.

tremely susceptible. Under gusty driving conditions, the man at the wheel is busy.

For those as optimistic as the speedometer on the test car, it is no trouble to push the needle to 110 MPH. The SPEED AGE staff arrived at a corrected and more conservative—but still fast enough—96 MPH. At that speed the Hornet's offspring is really working and the car feels, sounds and acts as though it were aptly named.

Having left innumerable Dragstrip Harry's shaking their heads in futility all across country to Washington, the test crew's appetite was whetted for timed acceleration runs. Zero to 60 MPH was turned repeatedly in 12.5 seconds and zero to 80 in 29 seconds—times that cannot be duplicated by many with much more muscle in the horsepower department.

Still convinced the trip-miles-per-gallon figures must be in error, the mileage meter was connected to check the test crew's arithmetic. This calls for driving at a constant speed, making sure the transmission has shifted into fourth gear, until the tenth of a gallon of gasoline in the visual bowl has been exhausted.

Making all possible allowances for human error, the results were still fantastic —31 miles per gallon at 30 MPH and 22 MPG at a steady 60. The dense traffic of metropolitan Washington made the Jet a bit more greedy—mileage dropped to 18.

Hudson built in the get-up-and-go characteristics and matched that performance when they designed the brakes. The 132 square inches of brake surface are more than adequate. Complete stops from 40 MPH were made repeatedly in an average of 71 feet and from 60 MPH in an average of 159 feet. Under repeated panic stops from 40 MPH, brake fade finally took over on the 18th such application of pedal—and that represents much more punishment than a braking system takes in days of driving.

Servicing of this car will present few problems to the mechanic. All the engine accessories are within reach although changing the plugs on those models equipped with Twin-H-Power almost requires the removal of the air cleaners.

The Jet differs from the Super Jet essentially in trim. Most of the extra-cost engine equipment can be ordered for the Jet also. Although now available only as a 4-door, 6-passenger sedan, there has been considerable speculation that Hudson will soon introduce a hardtop in the Jet line.

Safety-wise, the Jet must be rated highly. The remarkable lack of reflection-causing chrome is noteworthy and knobs on the dashboard have been held to a minimum. Visibility forward is excellent although the corner posts create the average blindspot. But their sturdiness and also that of the centerposts would probably go far to prevent a total roof collapse in a rollover. The combination of exceedingly rapid acceleration, high top speed, short wheelbase and fast steering, however, make the Jet an automobile to be treated with respect before attempting to master it.

Hudson's invasion of the low-price field, judging by the Jet, was well planned, for this only new car for '53 can do most of the fighting for them.

Road Test Report

Statistical Data

Engine Specifications

ENGINE:

- Cylinders 6
- Arrangement .... In-Line
- Valve arrangement .... L-head
- Bore (inches) 3
- Stroke (inches) 4.75
- Displacement (cubic inches) 202
- Compression ratio .... 8 to 1*
- Taxable horsepower 21.6
- Brake horsepower 114**
- Max. torque (foot pounds at RPM) .... 166 at 2000***
- Oil capacity (quarts) 5
- Fuel capacity (gallons) 15
- Water capacity (quarts)
  - without heater 15
  - with heater 16

Transmissions

Standard
Overdrive
Hydra-Matic

RATIOS:

	Hydra-Matic	Standard	Overdrive
1st.	3.8195	2.605	2.605
2nd.	2.6341	1.630	1.630
3rd.	1.450	1.000	1.000
4th.	1.000	——	.700
Reverse	4.3045	3.536	3.536

Interior Specifications

- Width of front seat at shoulder (inches) 54
- Width of rear seat at shoulder (inches) 55
- Depth of front seat (inches) 17⅜
- Depth of rear seat (inches) 18
- Headroom, front (inches) 36⅜
- Headroom, rear (inches) 35⅝
- Legroom, front (inches) 41⅞
- Legroom, rear (inches) 38½

Chassis

Frame:
- Type ..... Step down
- Wheelbase (inches) 105

Overall length (inches) 180 11/16
Overall width (inches) 67 1/16
Overall height (inches) 60⅞
Road clearance (inches) 68
Shipping Weight (pounds) 2,700
- Weight on front 55%
- Weight on rear 45%

Tread:
- Front (inches) 54
- Rear (inches) 52

Suspension:
- Front ..... Coil springs
- Rear ..... Semi-elliptic

Rear axle:
- Type ..... Semi floating
- Gearing ..... Hypoid
- Ratio 3.54****

Tires:
- Size ..... 6.40 x 15*****

Pressure:
- Front (pounds) 26
- Rear (pounds) 24

Brakes:
- Drum diameter (inches) 9
- Effective area (square inches) 132.14
- Type ..... Bendix

Steering:
- Type ..... Worm, roller
- Ratio ..... 20.2 to 1
- Turning diameter (feet) ..... N. A.

* Optional; standard is 7.5 to 1
** Optional head, Twin-H-Power; 106 with Twin-H-Power; 104 with 7.5 head
*** 158 at 1400 RPM with 7.5 head
**** Optional; standard transmission,
**** Optional; standard transmission, 4.1-3.31-4.27; Hydra-Matic transmission, 3.31
***** Jet, 5.90 x 15
N.A. Not available

Performance Data

Acceleration

3rd gear:
- 0-30 MPH ..... 3.9 seconds
- 0-40 MPH ..... 6.2 seconds

All gears:
- 0-30 MPH ..... 3.9 seconds
- 0-40 MPH ..... 6.2 seconds
- 0-50 MPH ..... 9.8 seconds
- 0-60 MPH ..... 12.5 seconds
- 0-80 MPH ..... 29.0 seconds

Top speed:
Average of two runs in opposite directions over measured mile, timed, 96 MPH

Braking:
- Complete stop 40 MPH ..... 71 feet
- 60 MPH ..... 159 feet

Speedometer Error:

Indicated	Actual
30 MPH	31 MPH
40 MPH	40 MPH
50 MPH	49 MPH
60 MPH	60 MPH
70 MPH	67 MPH
80 MPH	76 MPH
90 MPH	85 MPH

Fuel Consumption:
- 30 MPH ..... 31 MPG
- 40 MPH ..... 28 MPG
- 50 MPH ..... 24 MPG
- 60 MPH ..... 22 MPG
- Traffic ..... 18 MPG

PRICES

FACTORY

DELIVERED AT WASHINGTON, D.C.
INCLUDING FEDERAL TAX

- Jet $1,985
- Super Jet $2,085

Accessories; prices do not include state or local taxes:
- Hydra-Matic $178.03
- Overdrive $102.46
- Twin-H-Power ..... $ 85.60
- Radio, 8-tube ..... $ 99.82
- Radio, 6-tube ..... $ 81.81
- Heater ..... $ 72.72
- Tinted glass ..... $ 41.83
- Whitewall tires ..... $33.28 to $ 49.63
- Foam rubber cushions ..... $ 26.26
- Aluminum head ..... $ 11.74
- Wheel covers, rear ..... $ 14.54
- Wheel discs ..... $ 20.33
- Directional signals ..... $ 20.87
- Windshield washer ..... $ 11.17
- Electric clock ..... $ 21.94
- Back-up lights ..... $ 17.79
- Oil filter ..... $ 13.86

☆ ☆

HUDSON HORNET ROAD TEST

(*Continued from page* 37)

of dealing with cars which, above 60 m.p.h., entered terra incognita. There have been cars that ran 20 miles per hour slower than their speedometers at top speed. This seems hardly necessary, though the manufacturers might argue that it was an in-built safety factor.

ASR's road test crew found the Hornet a comfortable car, partly due to the great width of the seats and the ample leg and headroom. The ride was soft (though not overly soft and wind noise (as with earlier Hudsons) was reduced to a low level. The heating system worked uncbtrusively and well (which is what a heating system ought to do), though the heating fan was too audible at its "high" setting and even there could use a good deal more volume through the defroster vents.

Road noise was well-damped, though the engine could have been quieter at 70 and up. The trunk is not the largest in the industry, but contains enough space for any normal requirements—and the spare tire is mounted vertically inside it. The back seat has an enormous armrest that folds down from the center and gives an ample flat space for writing letters, map-reading, picknicking or what have you.

But the biggest comfort item in the Hornet, in our opinion, is the sense of security and relaxation that comes from its excellent road-holding, steering, and wheel adhesion—particularly at high speed.

There is no indication that the dual-carburetor Twin H power set-up gives more than a small percentage gain over Hornet single carburetor in normal road use—its chief advantage would appear to be in competition work where fractions count.

But the driver who demands safety with speed—particularly the long-distance, fast driver—can hardly overlook the Hudson Hornet when he's shopping around. ●●●

Hudson's new Jet can show its heels to the competition, and it should cost less to run. Will it be a

Threat to the Big Three?

A tendency to heel-over, as shown in the demonstration above, lessens the Jet's threat; but when that's corrected . . . watch out!

Photos by Jack Campbell

THE SALESMAN WILL tell you the Hudson Super Jet belongs in your garage. The Jet rates praise from the salesman because it's a very lively car, easy to drive, economical, and has a comfortable, soft ride. Its 114 horsepower rating (with Twin H-Power) is near the top of its field. It is priced to compete directly with the Big Three and the independents. Its wheelbase is shorter than other cars in the low-price field, so it has less room. Unless you're looking for spaciousness, you'll like the way the existing room has been divided between headroom and legroom, and front and back seats.

The car will appeal to many people. You may be the owner of a low-priced car, or a small car. You may have a large car and want a smaller one, either for a second car or to lower your expenses. If you fall into any of these groups, the Hudson Jet will interest you. MT Research analyzed the four-door Super Jet (equipped with Dual-Range Hydra-Matic and Twin H-Power) for this cross-section of buyers.

How does the Hudson Jet stand on economy? MT's Jet, with Dual-Range Hydra-Matic, averaged slightly better in overall mileage per gallon of Mobilgas regular fuel than many other cars in its class with automatic transmissions (see Performance Table). At high speeds (60 and 75 mph) the Jet gave mileages that are about average for low-priced clutchless cars.

Is the Hudson Jet powered to keep up with its competitors? Yes—and then some. MT's test car, with full Twin H-Power equipment (8.1:1 aluminum head and twin carburetors), proved to be a real jet in acceleration and speed checks. A standing start, quarter-mile time of 19.8 seconds placed it well out in front of its competitors. The Jet repeated this performance in its 0-60 mph runs, when it stopped the sweep-hand at 15.2 seconds, nearly six seconds faster than the average speed of others in its class. The car also proved it was well named by turning in a surprising 95.88-mph average top speed—again bettering the best in its class.

What gives the Jet its high-powered performance? Mainly, a scaled-down version of the H-145 Hudson Hornet engine, which gives more performance than is actually needed for a car of this size. The Jet gains a wider power range than any car in its class by offering a factory-supplied aluminum cylinder head and twin carburetors at an additional cost of $86. The standard Jet engine develops 104 horsepower, using a 7.5:1 iron head. With the optional aluminum head, the compression ratio is upped to 8.1, and the output is raised to 106 bhp. The aluminum head *and* the Twin H-Power dual carburetors boost the six-cylinder, L-head engine to 114 bhp. The small-bore, long-stroke (3 by 4¾ inches) six has a displacement of 202 cubic inches, much less than the Big Three class average of 230.9.

Does the Jet's powerplant crowd the engine compartment? Saying that the Jet's engine is a scaled-down Hornet engine applies to its size, as well as to its power, yet even though the Jet's six is no larger than similar engines in its class, its accessories are not easy to reach. The dual carburetor and twin air cleaner setup hides the spark plugs (this is true of many cars today that are using large air cleaners). The location of the fuel pump (under the exhaust manifold and behind the front crossmember) is poor; the pump can be removed or worked on only from beneath the car. The valves will also be hard to work on because of the location of the side-valve cover, mounted below the intake manifold in a hard-to-get-at place.

Is the Jet easy to drive? How does it handle under various conditions? A responsive steering system, which requires four turns of the wheel lock to lock, and a short overall length (slightly less than its class average) make the Super Jet an easy car to drive. It can be moved around quickly in traffic, it's extremely easy to park because of its compactness, and Hydra-Matic takes the strain out of stop-and-go driving. Maneuvering the car in tight places is no problem, for the car's overall width is just 67 inches.

Little steering correction is needed in tight corners, and at no time did the steering feel mushy or slow in reaction. The Jet had good response coming out of car tracks, and it wasn't necessary to fight the wheel on rutted roads.

Does it ride smoothly and quietly? Yes, it has an extremely easy ride. The corollary of such soft springing, of course, is greater lean on corners, and we feel that stiffer shocks would have produced a better compromise. The Jet reacts normally to average highway dips taken at speeds below 50 mph, but over that speed, oscillation (up-and-down body movement) is excessive. Bottoming, accompanied by three or four oscillations, occurred at 70 mph, and the front end felt nearly airborne. On sharp dips, full-spring travel was reached at 50 mph, and on very severe dips, the Jet bottomed at 30 mph. The Jet's recovery from dips is not rapid by comparison with other cars in its class, and it does not have the steadiness of its competitors. Excessive body vibration was felt only at very low speeds on washboard roads, and road noise was at a minimum under all conditions.

Is roadability good? The Hudson Jet has every reason to be proud of its heritage, but apparently the family did not include top-notch roadability as a hand-me-down: the Jet has few of the characteristics of the road-hugging Hornet. The Jet's cornering is not bad for a soft ride; the car reached a maximum point of body lean on MT's tight test curve at 40 mph. At 45 mph, the Jet's inside front tire became "light," and at 50 mph, it seemed nearly off the ground. On washboard roads, the Super Jet was quite stable, and it showed no signs of "fish-tailing" at speeds up to 50 mph.

Is Hydra-Matic a worthwhile addition to this car? The Jet offers an automatic transmission to pace the luxuries of its competitors. But since the Jet is an easy car to drive, we feel than an overdrive transmission is more adaptable to the Jet than a fully automatic transmission. The Hudson Jet can be equipped with Hydra-Matic for $178, or with overdrive for $102. As shown in the test tables, Dual-Range Hydra-Matic did not cripple the Jet's economy any more than it did its top speed, but an overdrive transmission should give the Jet even better economy figures, as well as reduce wear and tear on this high-piston-speed engine.

Does the Jet have good brakes? The Jet's brake mechanism is on a par with any car in its class. Effective lining area is adequate for a car of this size, and no abnormal lock-up or fade was encountered during MT's brake tests. However, the Jet's actual stopping power is not as good as other cars in its field. A check of the stopping distances of four other cars in the Jet's class showed distances that ranged from 189 to 205 feet (at 60 mph). The overall average of these cars (at speeds of 30, 45, and 60 mph) was 112. By comparison, the Jet stopped in 213 feet at 60 mph, and averaged 130 feet for 30, 45, and 60 mph.

Why is the Jet so hard to stop? The Jet's relatively poor showing is caused by a severe weight transfer that takes place when the front of the car noses down after the brakes are applied in a panic stop. During half the distance required to stop, the Jet "pulls down" nicely, keeping in a straight line; then the rear-end becomes light, throwing the car's weight forward. The result is a "snake dance," with the front end dipping excessively and the rear wanting to come around and meet the grille. Unless this condition is improved, it will be hard for the Jet to come up to the braking standard of its class.

Is the Super Jet roomy and comfortable? The shape of the seat does not lend itself to driver comfort, at least not as well as in other cars of this size. However, the passengers found the seats comfortable, shoulder room adequate, and headroom slightly better than average. Front seat legroom is good, but the Jet's size is reflected in "club coupe" rear seat legroom. Passengers in the rear seat had plenty of space for their toes under the front seat.

Does the Jet's short wheelbase limit the size of the trunk? The Jet has a rather large trunk for the size of the car. The interior has an average "finished" appearance, and only the floor is cloth-covered. The spare tire is carried on the right side of the compartment, in a vertical position. Although the 6.40 x 15 tire is comparatively small, and not too heavy, it still must be lifted to clear the six-inch sill at the rear of the trunk; this sill is necessarily high because of the placement of the gas filler pipe.

Are the Jet's interior appointments good? A recount of "first impressions" will probably include mention of the appearance of the Jet's interior, for a first glance reveals a nicely finished, pleasing combination of design, color, and good-looking materials. The foam rubber-padded seats are covered with a soft fabric which we considered excellent for this price class. The back of the front seat has a large pocket for storage of odds and ends, and a big padded roll at the top can add to rear passenger safety. The lower door panel sections and the large, built-in armrests (an integral part of the door), are trimmed with leather-like plastic; interior trim is complemented by bright chrome hardware.

From the seat padding to the rubber floormats, the Jet's interior rates an "excellent" for workmanship and quality of

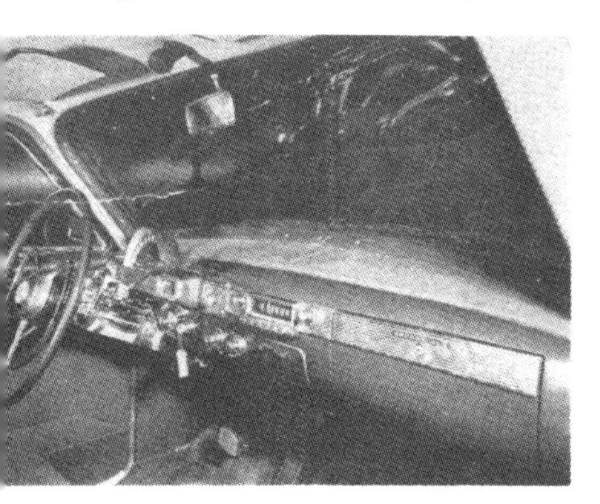

Chrome and color adorn the Jet's dash, but the plastic-topped panel stops glare

The Jet's back seat is attractive and well padded, has the traditional Hudson pocket

Trunk is roomy for a car of this short wheelbase. Gasoline filler is near center

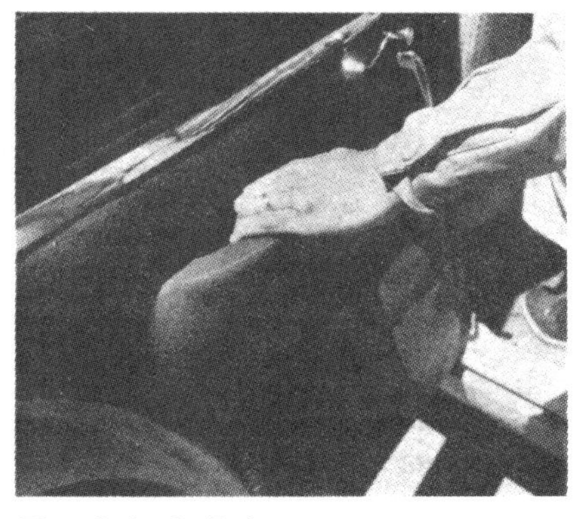

The Jet's built-in armrests have large hand grips, are comfortable and practical

material. Still, its usefulness will be limited to short trips with a full passenger load, or to small families, because of the lack of space for rear seat passengers.

How safe is the Hudson Jet's interior? Like many cars, the Jet has some features that are commendable, as well as some that deserve mention as drawbacks to safety. A flush-mounted pushbutton on the glove compartment door eliminates projections on the right side of the dashboard, but the center of the panel is well studded with knobs, the ashtray handle, and a protruding radio control panel. The Jet deserves praise for the use of glare-reducing plastic on the top of the instrument panel. The plastic covering is much more effective than dull paint, for there is no glare from its surface under any condition. The dashboard ashtray is within the driver's reach. The glove compartment, a conventional type of average size, is not particularly handy for the driver.

Are the Jet's instruments easy to read? Are the controls well marked and easy to use? The Hudson Jet's large, hooded speedometer gives the driver the impression of sitting behind the wheel of a much larger car, and its position high on the dashboard allows him to read it without lowering his eyes too far from a normal view of the road. Fuel and temperature gauges and the generator and oil pressure warning lights are mounted just below and to each side of the speedometer. All instruments have excellent readability because their numerals and indicators are mounted on a dull black background. The chrome headlight, lighter, and windshield wiper control knobs are not too plainly marked, but their position should keep confusion at a minimum. The Dual-Range Hydra-Matic selector dial is lit for night operation. Heat and vent controls are to the left of the steering column. They are not lit for night use, but their operation is simple, and the controls are within easy reach of the driver. The Jet's T-handled emergency brake is under the left side of the dash panel; it is easy to pull on and release.

Even without the optional twin carburetors, engine accessibility is not up to par

Does the driver have maximum visibility in the Jet? The one-piece windshield of the Jet does allow good all-around visibility. Blind spots are eliminated by the narrow corner posts in front and rear and the one-piece, wrap-around rear window.

The rear-view mirror is high enough to be out of the driver's line of vision to the right front, and the combination of fairly high seats and a slightly sloping hood allows a good view of the road. The wipers should clear a greater area than they do.

The driver can see the right front fender without sitting up high and peering over the hood, but he won't see the right rear fender without getting out of the car.

Does the heat and vent system work well? The Jet's heating system is excellent. Its output should be entirely satisfactory for cold-weather driving, and the temperature of the warm air is very easily controlled. In warm weather, the temperature control is moved to the "Summer—Off" position, and fresh air comes into the car through the cowl ventilator.

Is the Jet a well-made car? "Step-down" design, in which the frame rails are above the floor level, has been retained in Hudson's new car. The body panels, grille, and bumpers are of good-quality metal, and the gauge is heavy enough for the job. Integral rear fenders may add to repair costs, being a part of the rear quarter-section of the body. You might do well to put touch-up paint down on your list of "things to buy for my new Jet," for there is no protective metal strip along the fenders or doors. Although this leaves the panels susceptible to scratches and dents from doors being bumped against them in crowded parking lots, the absence of chrome adds refinement to the car. The Jet's fenders are tipped with large, cone-shaped taillights, which can be seen from the sides of the car at night. Wheel openings are high enough to allow easy tire removal with the bumper jack supplied with the car.

The hood and rear deck lid have over-center springs for easy opening. The hood, released by an under-grille catch, is well braced, and stays open with no likelihood of slamming itself down on an unsuspecting mechanic's neck. The rear deck lid, also firmly supported, is opened easily by a key-operated latch.

The Jet's doors swing wide to allow easy entrance and exit for front seat passengers. Getting in and out of the rear compartment is somewhat complicated by the far-forward position of the rear seat. Door construction is good, and all doors are fitted with a stop. Except for an ill-fitting left-rear door, all body panels on the MT test car were faired smoothly, and no uneven gaps were seen. The Jet's finish was good, and showed no blemishes or traces of "orange peel."

Will the Hudson Jet be a durable car? With its "Monobilt" body and chassis (welded together to form a single unit), the Jet should give many serviceable, trouble-free miles. The only annoying squeaks and rattles in our test car came from the front bumper, which eventually parted from the right-hand bracket. However, this was not a failure of body and frame construction and was easily remedied. Halfway through the road test, the Jet developed a miss and ran somewhat roughly, but this was corrected by a much-needed tune-up. The Hydra-Matic unit and all running gear remained quiet and untroubled, and the car's interior was in showroom condition at the end of the test.

Will the Hudson Jet be a real threat to the leaders of the low-price class? The Jet may make a dent in certain categories of the low-price field, for it is much more impressive (inside and out) than certain non-deluxe versions offered by the competition. These undecorated models, long respected by the driving public and by MOTOR TREND, offer perfectly satisfactory transportation at lower cost. However, manufacturers and dealers have made standard models hard to obtain.

To be a real threat, the Jet will need more than its excellent performance to make that long, uphill grind to the top of the low-price field. Its engine needs few mechanical improvements, and its body lines, typical of its class, need no period of introduction to the car buyers. A wider tread plus heavier shocks would improve stability. Increased rear passenger space (which could be brought about in conjunction with improved weight distribution) would make the Jet a stronger challenger to the leaders in its chosen price class.

Who will buy the Hudson Jet? The Jet's biggest edge over the competition lies in its outstanding performance. It is full of pep, and it gives good gas mileage. People who are looking primarily for those features in a low-priced car will go for the Jet in a big way. Furthermore, they will find it ideal for around-the-town trips, and they'll be able to park the Jet in spaces that they would have to pass by with larger cars.

The Hudson Jet is the only all-new car to be introduced this year. We believe that widespread acceptance of the Jet may take some time. Why? Because the Jet, rather than being a new car in this field, is more a new *kind* of car in the low-price class. The next two years will tell whether buyers will respond *en masse* to its features, or whether they prefer a more conventional package.

(For The Story in Figures, see page 47)

THE STORY IN FIGURES

1953 HUDSON SUPER JET

(Equipped with Dual-Range Hydra-Matic and Twin H-Power)

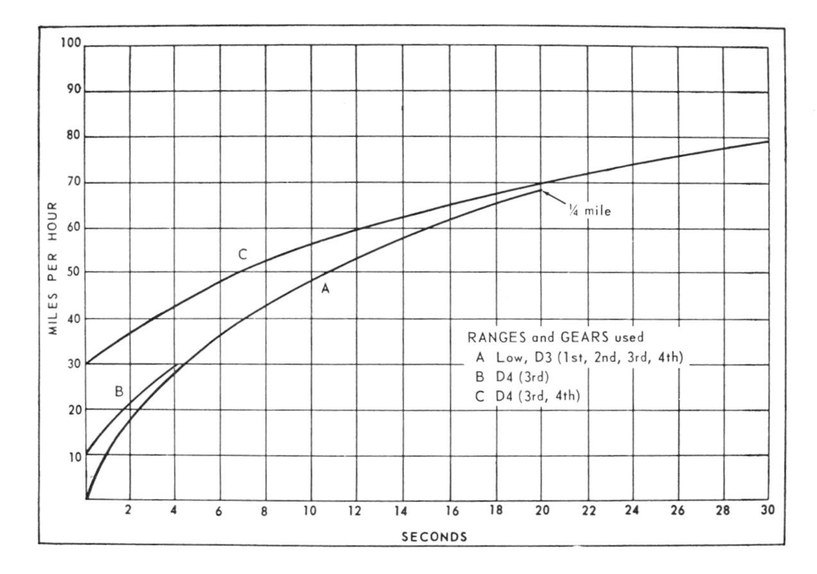

ACCELERATION

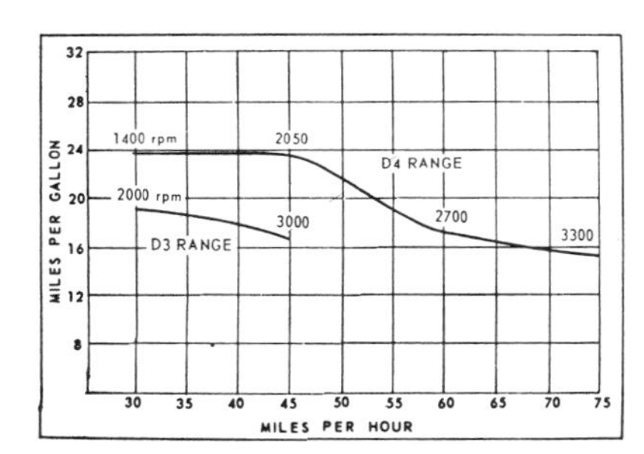

FUEL CONSUMPTION

As the Jet is Hudson's first low-priced car in many years and is still in its first year, no reliable depreciation figures are available.

PERFORMANCE

CHASSIS DYNAMOMETER TEST

(Checked on Clayton Mfg. Co.'s chassis dynamometer; all tests are made under full load conditions)

RPM	MPH	ROAD HP
1200	27	28
2000	45	48
2500	57	62
3200	73	70
3500 (maximum)	80	73

ACCELERATION

(In seconds; checked with fifth wheel and electric speedometer)

Standing start ¼ mile (68.6 mph; LOW and D-3 range)	19.8
0-30 mph (0-29, car speedometer; LOW and D-3 range)	4.6
0-60 mph (0-56, car speedometer; LOW and D-3 range)	15.2
10-20 mph (DRIVE range)	1.7
20-30 mph (DRIVE range)	2.4
30-40 mph (DRIVE range)	3.1
40-50 mph (DRIVE range)	3.6
50-60 mph (DRIVE range)	6.0
60-70 mph (DRIVE range)	7.2
70-80 mph (DRIVE range)	11.9

TOP SPEED

(In miles per hour; clocked speeds over surveyed ¼ mile)

Fastest one-way run	102.38
Slowest one-way run	89.02
Average of four runs	95.88

FUEL CONSUMPTION

(In miles per gallon; checked with fuel flowmeter, fifth wheel, and electric speedometer)

Steady 30 mph	23.4
Steady 45 mph	23.3
Steady 60 mph	16.8
Steady 75 mph	14.8
Traffic	16.4

BRAKE STOPPING DISTANCE

(At speeds shown; checked with electrically actuated detonator)

30 mph	45 ft.
45 mph	104 ft.
60 mph	213 ft.

GENERAL SPECIFICATIONS

ENGINE

Type	L-head, six cylinder
Bore & stroke	3.00 x 4.75
Stroke/bore ratio	1.58:1
Compression ratio	8.0:1
Displacement	202 cu. in.
Advertised bhp	114 @ 4000 rpm
Piston travel @ max. bhp	3166 ft. per min.
Bhp per cu. in.	.564
Maximum torque	166 @ 2000 rpm
Maximum bmep	123.92 psi

DRIVE SYSTEM

Standard transmission	Three-speed synchromesh using helical gears
Ratios	1st 2.60, 2nd 1.63, 3rd 1.1, reverse 3.53
Automatic transmission	Hydra-Matic, fluid coupling with gears
Ratios	1st 3.82, 2nd 2.63, 3rd 1.45, 4th 1.1, reverse 4.30
Overdrive transmission	Planetary type with manual lockout and accelerator downshift
Ratio	.7:1 (overall 2.98)
Rear axle ratios	Conventional, 4.1 standard, 4.27 and 3.31 optional; Hydra-Matic, 3.54 standard, 3.31 optional; overdrive, 4.27 standard, 4.1, 3.54 and 3.31 optional

DIMENSIONS

Wheelbase	105 in.
Tread	Front 54 in., rear 52 in.
Wheelbase/tread ratio	1.98:1
Overall width	67 in.
Overall length	186.8 in.
Overall height	60.3 in.
Turning diameter	36 ft.
Turns lock to lock	4
Weight (test car)	3050 lbs.
Weight/bhp ratio	20.75:1
Weight distribution	Front 56.7%, rear 43.3%
Weight/sq. in. brake lining	23.1 lbs.
Tire size	6.40x15
Tire loading (% of recommended maximum at curb weight)	Front 89.6%, rear 68.4%

SAFETY CHECK

DRIVER SAFETY:

	YES	NO
Blind spot at left windshield post at a minimum?	X	
Blind spot at rear vision mirror at a minimum?	X	
Vision to right rear satisfactory?	X	
Windshield free from objectionable reflections at night?	X	
Dash free of annoying reflections?	X	
Left side of dash free of low projections?		X
Cigarette lighter, ashtray and glove compartment convenient for driver?		X

DRIVER AND PASSENGER:

	YES	NO
Front seat apparently locked securely at all adjustment points?	X	
Metal strip eliminated between front quarter window and main door window?		X
Rear view mirror free of sharp corners?	X	
Right side of dash free of projections?	X	
Adequate shock-absorbing crash pad?		X

REAR SEAT PASSENGERS:

	YES	NO
Back of front seat free of sharp edges and projections?	X	
Rear interior door handles inoperative when locked?	X	
Adequate partition to keep trunk contents out of passenger compartment on impact?		X

PRICES

(Including retail price at main factory, provisions for federal tax, and delivery and handling charges, but not freight.)

	Jet	Super Jet
Four-door sedan	$1858.00	$1954.00

ACCESSORIES

Dual-Range Hydra-Matic	$178.03
Overdrive	102.46
Radio	81.81
Heater	72.72
White sidewall tires (additional cost per set, 6.40 x 15)	34.06
Tinted glass	41.83
Twin H-power with 8.1:1 alum. head	85.60
Turn indicator	20.87
Wire wheels	123.63
Continental kit	169.50

OPERATING COST PER MILE

(In this portion of the test table, MOTOR TREND includes those items that can be figured with reasonable accuracy on a comparative basis. The costs given here are not intended as an absolute guide to the cost of operating a particular make of car, or a particular car within that make. Depreciation is not included.)

Cost of gasoline	$132.75
Cost of insurance	116.60
Maintenance:	
Wheel alignment	7.20
Brake reline (front only)	17.30
Major tuneup	13.75

(Labor only; includes: clean and adjust or renew points; disassamble and clean carburetor; adjust spark timing, carburetor, valves, fan belt and generator; clean air cleaner, battery terminals and fuel lines; check coil, condenser, vacuum control, heat control, compression and vacuum; tighten cylinder head, manifolds and hose connections.)

Automatic transmission (adjust, change lubricant)	8.30
First year operating cost per mile (based on 10,000-mile annual average)	3.0c

MAINTENANCE AND REPAIR COST ANALYSIS

(These are prices for parts and labor required in various repairs and replacements. Your car may require all of them in a short time, or it may require none. However, a comparison of prices for these sample operations in various makes is often of pertinent interest to prospective owners.)

	COST	LABOR
Distributor	$25.45	$ 2.40
Battery	22.05	1.05
Fuel pump	6.05	2.40
Fan belt	2.25	1.20
Valve grind	3.15	25.60
One front fender	35.00	12.50
Two tires	44.02	
One bumper	30.00	3.00
TOTALS	$167.97	$48.15

The Hudson Line for '54

INCREASED horsepower highlights the Hudson line of 1954 cars, first in the new car parade, which were formally announced October 2. The addition of the Jet-Liner to the Jet series is another feature.

All series—Hudson Hornet, Wasp and Super Wasp, Jet, Super Jet and Jet-Liner—boast new performance as a result of Hudson's newly developed "Instant Action" engines with Super Induction. The company describes the latter as "an improved induction system which increases combustion efficiency and provides quicker response at every point in the driving range."

"Flight-Line Styling" of all models makes them appear lower and longer viewed from any angle. Interior appointments are declared more luxurious than in any models in company history.

The "step-down" design with Monobilt body-and-frame construction is continued in all models.

Hudson power steering—for effortless parking and steering control at all speeds—and improved power brakes are available as optional equipment for the Hornets, Wasps and Super Wasps for 1954.

The company declares the introduction of "Instant Action" engines "sets new standards of power and performance for high-compression engines." Super Induction provides an instant surge of power at a touch of the accelerator, providing an added margin of safety for emergencies.

Super Induction results from an ingenious departure from ordinary combustion chamber design, whereby the gas-air fuel mixture is used with greater efficiency, the company states. The redesigned combustion chamber, together with a silent Long-Dwell Camshaft, increases power production at all speeds, even at higher ranges where reserve engine power ordinarily begins to diminish.

1954 HUDSON HORNET offers "Flight-Line Styling,"—appears lower and longer. All new Hudsons have "Instant Action" engines with "Super Induction."

For 1954 Hudson hood lines blend into functional air scoop, which provides better engine ventilation. In the newly designed grille the Hudson emblem is set in the middle of the horizontal center bar. Frontal lines are designed to give a wider, lower-built appearance.

As to body trim, highly polished rub rails, located below the side panel flair, extend from front to rear. Panels extend from the rub rails to the lower edge of the rear fender, with a shorter rub rail running parallel to the top rail over the rear wheel opening.

Huge, triangular tail lights are visible from side as well as rear, being set high in the jet-stream rear fenders, which give an impression of increased car length. The trunk lid on the Hornet and Wasp models has been redesigned for styling and added trunk room. Massive, profile-type rear bumpers wrap around the edges of rear fenders, giving added protection and increased strength. A "built-in" license plate frame in the rear bumper also acts as a center guard.

In the interiors long wearing upholstery fabrics in two-tone color combinations are harmonized with exterior car colors. Door panels are recessed for added elbow room and trimmed in two-tone materials. Door pulls and window controls are placed in the recessed doors.

Instrument panels have instruments grouped directly before the driver for instant viewing. Dials are illuminated by indirect lighting. Teleflash signals flash an immediate warning if oil pressure or generator charging rate drop.

All '54 Hudsons have huge, one-piece curved windshields with no center bar to obstruct vision. Wrap-around rear windows and elongated corner

INTERIOR of the Hornet is trimmed in Plasti-Hide. Instruments are grouped in a unit directly in front of the driver. All '54 Hudsons have one-piece curved windshields.

windows provide extra rear-view safety.

Hudson power steering is the direct-action, linkage type. It takes over when the steering wheel is turned as little as 2 degrees, providing as much as 80 per cent of required steering effort.

Spring resistance of four pounds gives a constant "feel of the road" and prevents dangerous oversteering.

As an added safety factor, Hudson Power Steering is designed with the same steering ratio and linkage as Hudson's Standard Center-Point steering system. Even should the power assist fail, as it might in any car due to accident, the car can still be steered manually, and without the severe extra effort required by other systems when power fails.

In this power steering unit no seasonal oil change is required, working parts being permanently sealed in oil.

Hudson's Power Brake system is a combined vacuum and hydraulic unit

NEW ADDITION to low-priced line is the Jet-Liner with 104-horsepower engine. It uses Hudson "Mono-built" construction.

New Styling for Hornet and Wasp — Jet Liner Added — More Power in All Models

operated by a pendulum type foot pedal.

Braking effort is reduced by 65 per cent. With power brake pedal on the same level with accelerator, the driver merely pivots his foot without lifting his heel from the floor to produce instant and almost effortless braking action. The power brake pedal travels only one-sixth the distance of the pedal travel of a conventional braking system, greatly reducing brake reaction time for emergency stops.

Hudson's Power Brakes provide two extra margins of safety if, due to accident, the power source should fail. A reserve vacuum tank permits up to three power brake applications. In addition, a direct mechanical linkage operates the brakes hydraulically without power assistance. The hydraulic system is equipped with another safety feature; a reserve hydraulic fluid tank that keeps the hydraulic system filled to a safe level at all times.

The Hudson Power Brake system has a permanently lubricated unit designed for years of trouble-free service. All working parts are sealed in oil. The unit is easily accessible for service. Under ordinary conditions, only regular maintenance checks of the hydraulic system itself is necessary.

Pacing the 1954 Hudson line is the Hudson Hornet, national stock car champion, which holds every national A.A.A. stock car record.

Its new "Instant Action" engine is rated at 160 horsepower, with compression ratio of 7.5 to 1. With Twin H-Power, Hudson's popular multiple fueling system available at extra cost, the Hornet engine delivers 170 horsepower. Engine displacement is 308 cu. in. Bore is 3-13/16 inches and stroke, 4½ inches. Aluminum cylinder head is standard equipment.

The Hornet's L-head, in-line engine delivers peak performance on regular gasoline, premium-priced fuels not being required. As in all '54 Hudsons, the Hornet engine offers super-hard chrome alloy cylinder blocks.

The Hornet's lower-priced running mates—the Hudson Wasp and Super Wasp—offer many of the Hornet's advantages in the medium-price field.

With standard cast-iron cylinder head, the Super Wasp "Instant Action" engine develops 140 horsepower and has a compression ratio of 7 to 1. With optional aluminum head, it has 143 horsepower, 7.5 to 1 compression ratio. With Twin H-Power and aluminum head, horsepower is boosted to 149. Engine displacement is 262 cu. in.

The Wasp's "Instant Action" engine has a compression ratio of 7 to 1 with cast-iron head and develops 126 horsepower. With optional aluminum head, compression ratio is 7.5 to 1, horsepower 129. Engine displacement is 232 cu. in.

Twin H-Power, Hudson's new multiple fueling system chosen by more than 50 per cent of Hudson buyers since its introduction, is available as optional equipment for all models except the Wasp. Twin H-Power is said to accurately measure and distribute gasoline to each cylinder.

Dual-Range Hydra-Matic Drive, when available, is offered on all '54 Hudsons as optional equipment. Overdrive is optional on manual shift, syn-

REAR of Hornet and Wasp has new trunk contour with more luggage space. High-mounted tail lights make for longer look.

cro-mesh transmissions.

Redesigned trunk lids provide additional luggage space in the new Hudsons. Usable trunk room has been increased by use of new-type thin hinges placed at the extreme edges of the trunk lid. This also serves to prevent damage to luggage. Spare tire is located in upright position for easy removal and more usable space.

The '54 Hornets, Wasps and Super Wasps are offered in a wide range of sparkling new solid colors and two-tone combinations.

The Hudson Hornet series is available in a four-door sedan, club coupe, Hollywood Hardtop, and convertible brougham.

The Hudson Super Wasp is offered as a four-door sedan, club sedan, club coupe, Hollywood Hardtop, and convertible brougham.

The Hudson Wasp is offered as a four-door sedan, club sedan and club coupe.

The Hudson Hornet and Super Wasp are available in 12 solid colors and 12 two-tone combinations. The Wasp offers a choice of 12 solid colors.

All '54 Jets have the newly developed "Instant Action" engines with Super Induction, and the highest power to weight ratio in the light car field is claimed for them.

The new Jets are built lower. Interiors are roomier with rear seat passengers having more than two inches of added leg room. Interiors are more luxurious.

All '54 Jets have "step-down" design and lower center of gravity.

All Jet engines have oversize bearings and rugged construction of the Hornet's engine. Cylinder blocks are of chrome alloy for long motor life and minimum oil consumption. Due to its simple design, the Jet's in-line, L-head engine has fewer moving parts and wear points, which means lower upkeep costs.

The Jet's "Instant Action" engine has a compression ratio of 7.5 to 1 with cast-iron cylinder head, and 8 to 1 with optional aluminum head. Premium fuel is not required. Bore is 3 inches and stroke 4¾ inches. Displacement is 202 cu. inches. The Jet engine develops from 104 to 114 horsepower, depending on choice of equipment.

Twin H-Power is available as extra cost equipment for all '54 Jets.

The Jet has a full-size suspension system with airplane-type double-acting shock absorbers in both front and rear. It has independent front wheel coil springing of special alloy steel for smooth, easy riding on the roughest roads. Dual-action front stabilizer bar and diagonally mounted rear shock absorbers, together with splay-mounted rear springs, control sway and provide stability on the sharpest curves, it is claimed.

All Jets have center-point steering which acts directly and equally on both front wheels, thus eliminating "wheel fight" and road wander.

New addition to the Hudson Jet series is the Jet-Liner, which the company boasts brings new concepts of luxury and fine craftsmanship to the low-price field.

Interior of the Jet-Liner features pleated seats, cushioned with foam rubber, and an interior top upholstered in antique-white Plasti-Hide.

The lowest-priced Hudson Jet, formerly available as a four-door sedan, is also offered as a six-passenger Utility Sedan in the '54 series. This practical two-door Jet is designed for family, farm or business use. With the back seat in place, it is a roomy, six-passenger sedan. With back seat removed, the trunk divider wall falls forward to offer a spacious area for carrying tools, sports equipment, salesmen's samples, or extra luggage.

A wide range of gear ratios is offered to suit every Jet owner. The buyer may select the ratio best suited for his individual driving requirements—and at no extra cost. Standard gear ratio with manual transmission is 4.10 to 1. For cars operating in mountainous territory, a special gear ratio of 4.27 to 1 is offered. With Dual-Range Hydra-Matic, gear ratios of 3.54 to 1 and 3.31 to 1 are available.

Body styles include the four-door sedan and club sedan Jet-Liner; the Super Jet in choice of four-door sedan or club sedan, and the Jet as a four-door sedan or two-door Utility Sedan.

All models offer a wide choice of new colors. The Jet is available in six standard, two optional solid colors and four special colors. The Super Jet comes in a choice of eight standard, four special solid colors and twelve two-tone combinations. The Jet-Liner is available in eight standard, four special solid colors and twelve two-tone combinations.

★

THESE ARE THE INDUSTRY-LEADING ADVANTAGES THAT ENABLE HUDSON ALONE TO BRING YOU

"The New Step-Down Ride"!

Exclusive "step-down" design with its recessed floor • The steady, road-hugging security of the lowest center of gravity in any car • Added safety and quietness of Monobilt body-and-frame* • Roomiest seats in any mass-produced car built today • Low-built, free-flowing lines of true streamlining • Amazing head room in both front and rear seats • Full road clearance • Curved Full-View windshield • New High-compression Pacemaker Engine • High-compression Super-Six Engine, America's most powerful Six • Or even more powerful Super-Eight • Exciting Super-matic Drive†—only automatic transmission that gives you conventional shifting at the touch of a button, and combines fuel-saving overdrive • Powerful Triple-Safe Brakes • True Center-Point Steering • Super-smooth Fluid-Cushioned clutch • Super-Cushion tires • Chrome-alloy cylinder block • and more than 100 other outstanding improvements.

*Trade-mark and patents pending. †Optional at extra cost

HUDSON TESTS A SPORTS CAR

New Hudson's grille, above, is still in formulative stage, may be changed in final presentation of model. Familiar Hudson "frown" will be retained, however. License plate is deeply recessed, right, lighted from within.

Parking lights are set back into sides of front fenders, form tips of chrome spears.

Hot from the Touring body factory in Italy comes the latest entrant in America's glamour-buggy derby—the Hudson Super Jet. Still so new that performance data is unavailable, it has crossed the ocean loaded with innovations.

Vital statistics include stock Jet chassis and power plant, a body mostly of aluminum, wire wheels and strictly American-type massive bumpers. Seats have safety belts.

Hudson's immediate plan is to get public reaction to this Superleggera-bodied coupe. If favorable, the company may go into limited production with a larger engine and a "competition kit." • —Walter Von Schoenfeld

Triple "exhausts" on rear fenders, house the stop lights, taillights and backup lights.

Doors have been cut into roof to allow easy entrance into low-slung, step-down interior.

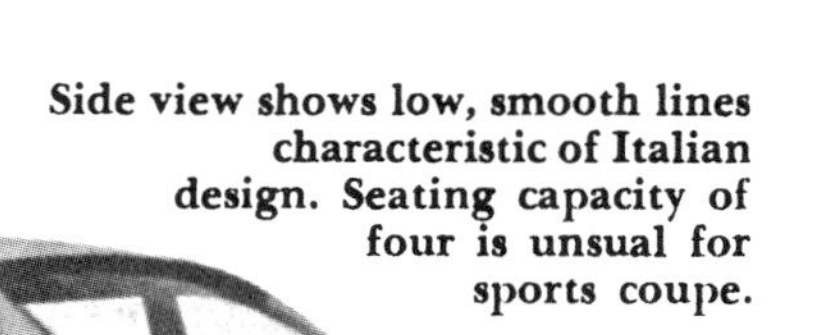

Side view shows low, smooth lines characteristic of Italian design. Seating capacity of four is unusual for sports coupe.

Photos by Ralph Poole

Road Test

HUDSON HORNET

BY ART NICHOLAS

WHAT does Hudson have to offer in its Hornet model for '54? This is an important question to the stock car racing fan as well as the average motorist.

The reputation of the company, for decades, has been solidly built by record-smashing automobiles. In fact, of all Detroit manufacturers, Hudson is the one which openly has been the most competition-minded for the longest period. And, it should be unnecessary to add, with the greatest success both on the track and the measured straightaway. Some of the marks its cars have set as far back as the late twenties and throughout the thirties still stand.

Recent performance of the Hornet in stock car competition is almost too well-known to repeat. Let's simply say that it has been the car to beat in all NASCAR and AAA stock car events. The times when it has not collected the checkered flag have been exceedingly rare.

This was a car that the MOTOR LIFE road crew tested with respect. The particular model of the Hornet chosen carried all the refinements skilled Hudson engineers have made to date, including Super-Induction, Twin H-Power, power steering and power brakes.

ACCELERATION

In comparison with the huge ohv V-8 engines that are becoming the rule in the big-car field, the Hornet's output of 170 hp is almost modest. However, the dependable L-head shows up best in the lower ranges where its good torque is most effective. This can be noted in the 0-30 mph figure, which was an exceptional 4.2 seconds. Tests made from a standing start to 45 and 60 mph were not quite so remarkable but will stand up favorably against those earned by any U.S. stock car on the market today.

TOP SPEED

Full throttle runs through the measured quarter-mile, half-mile and mile were made with an 11 mph sidewind. Nevertheless, in this department, the Hornet showed the greatest improvement in performance over 1953. The average time was 106.54 mph, although on one quick trip through the traps the Hornet was clocked at 108.43 mph. The tests were made, incidentally, at the favorable altitude of 430 feet above sea level.

MILEAGE

Equipped with Dual-Range Hydra-Matic, the Hornet's gasoline consumption was metered at 17.1 mpg at a steady 30 mph down to 14.2 mpg at a steady 60. The overall average for the entire test period, which included 1,011 miles of mountains, flat highway and city traffic, was 15.5 mpg.

ENGINE

The Hornet can be obtained with a variety of options in its engine. Among the recent improvements made are Twin H-Power (dual carburetion) and Super Induction (refinements in the valve layout for better breathing and a hotter cam). These are the chief reasons for the Hornet's increased performance. Other basic good points are inherent in the design which was introduced early in the postwar period. Although its sound could be toned down and the oil dip stick located a bit more conveniently, the overall arrangement in the forward compartment is good, as an inline engine can be expected to be, despite the loading of power-operated accessories.

TRANSMISSION

The Hornet tested was equipped with the familiar and popular Hydra-Matic. This, however, will not be true of most other '54 Hornets. Company engineers told the road crew that later models in the year will carry the Borg-Warner unit, an item which they opined would improve the performance characteristics of the car. The switch probably has something to do with the fire last summer at the Hydra-Matic plant, although other factors undoubtedly influenced Hudson's decision.

HANDLING

This is the department that causes the Hornet to be rated extremely high by the test drivers, perhaps even more than the performance. The oft-advertised "step-down" design of the frame does make a great contribution to the over-the-road control of the car. Corners, which many other cars would take with excessive body sway, are a pushover for the Hudson. To further experiment with this feature, drivers took the car over an exceedingly difficult paved mountain road. Sharp curves on up and down grades were no problem and almost a pleasure. The stability of the car and the new power steering are a perfect team.

A special note should be added here about Hudson's power steering. Test drivers, frankly, are not too happy about the innovation—on most cars. All power steering devices are helpful on slow corners, in traffic, and while parking, but leave something to be desired at speed on the open road. Not so with Hudson's. It has plenty of feel and is just easy enough to take the work out of driving and retain the fun. During the speed runs at more than 100 mph with a sidewind, the car handled easily and held to the road exceptionally well.

The power brakes, as an item of control, are difficult to classify. They are there, but more pedal pressure is needed than with other types. Whether this is good or bad depends upon the individual. The important difference is that slightest movement of the foot does not produce the sudden, tire-burning halt. The driver still has to do the braking.

RIDE

The same factors which give the Hudson favorable handling characteristics contribute to its good ride. Like any heavy car, it is smooth and bounce freee. In addition, there is much less rolling and pitching. Driver and passengers ride low in the car and feel it.

(Continued on page 54)

GENERAL SPECIFICATIONS

MAKE & MODEL: 1954 four-door Hudson Hornet
ENGINE: six-cylinder inline L-head
HORSEPOWER: 170 @ 3,800 rpm
DISPLACEMENT: 308 cubic inches or 5,055 cubic centimeters
COMPRESSION RATIO: 7.5 to 1
BORE & STROKE: 3 13/16 x 4½
MAXIMUM TORQUE: 278 ft.-lbs. @ 2600 rpm
TRANSMISSION: Hydra-Matic
REAR AXLE RATIO: 3.58 (with standard transmission, 4.09; with overdrive, 4.55)
WHEELBASE: 124 inches
TREAD: 58½ front, 55⅜ rear
OVERALL LENGTH: 208 5/32
OVERALL WIDTH: 77.6 inches
OVERALL HEIGHT: 60.37 inches
ELECTRICAL SYSTEM: six volt
TIRES: 7.10 x 15
FUEL CAPACITY: 20 gallons
TEST WEIGHT: 3,970 lbs.
POWER/WEIGHT RATIO: 21
STEERING—
Turns, Lock-to-Lock: 5½
Turning Circle: 32 feet, 6 inches

PRICES

(The figures quoted are the advertised delivered retail list prices as suggested by the factory. They include Federal taxes, recommended delivery and handling charges. They do not cover transportation costs, state or local taxes, optional equipment or any other charges that may be made by a dealer.)

FOUR-DOOR: $2,768.86
CLUB COUPE: $2,741.99
HARDTOP: $2,987.75
CONVERTIBLE: $3,287.70
ACCESSORIES—
Automatic Transmission: $178.03
Overdrive: $110
Power Steering: $177
Power Brakes: $48
Radio: six-tube, $68; eight-tube, $100
Heater: $75
Directional Signals: $22
White Sidewalls: $48 extra per set
Wire Wheels: $290.25 chrome, $123.63 painted for set of five

TEST CONDITIONS

WEATHER: clear
TEMPERATURE: 84 degrees
WIND: 11 mph, right angles
ALTITUDE: 430 feet above sea level
MILEAGE AT START: 3,960
MILES COVERED: 1,011
GASOLINE: 91 octane
OIL: 30 weight
EQUIPMENT: electric fifth wheel, electric speedometer, electric odometer, Fuel-flow meter, Perfometer, calibrated clocks
SPEED TEST SURFACE: asphalt
SPEED COURSE: surveyed quarter-mile, half-mile, and mile

PERFORMANCE

ACCELERATION IN SECONDS—
Standing Quarter-mile: 18.5
0-30 mph: 4.2
0-45 mph: 8.9
0-60 mph: 12.2
TOP SPEED—
Fastest One-Way Run: 108.43 mph
Slowest One-Woy Run: 104.61 mph
Average of Six Runs: 106.54 mph
FUEL CONSUMPTION—
Steady 30 mph: 17.1 mpg
Steady 45 mph: 16.3 mpg
Steady 60 mph: 14.2 mpg
Total Test Mileage: 1,011 miles @ 15.5 mpg

Hudson appearance has been brought up to date with sweeping fender line at rear and new contour of the hood

Careful checks with accurate electric fifth wheel showed the Hornet still packs plenty of performance and remains a potent stock car racing threat

Member of the road test crew illustrates the small amount of foot travel necessary between the Hornet's accelerator and brake pedals, a highly desirable feature

Massive chromed bumpers hug the Hudson Hornet body closely and eliminate the traditional location of the gravel pan

HUDSON ROAD TEST

(*Continued from page* 53)

EXTERIOR

In body styling, the most obvious change on the Hudson is in the rear fender line, which has been raised and lengthened to give the car a longer, more streamlined look. The comment of the MOTOR LIFE staff photographer, who has put countless cars on film, was interesting: "It sure looks sexy," he said.

Contour of the hood, which could stand strengthening, has been rounded. Other modifications are chiefly in trim and grille decorative touches.

INTERIOR

Entry into the Hudson Hornet is easy, with doors opening at an 80-degree angle. Alterations in the interior for the '54 are few. There still are numerous lights for the convenience of passengers: two in back, two in front and one overhead. The back of the front seat continues to carry the almost extinct robe rope. Arm rests, both front and rear, are not located in the most comfortable positions. It's interesting to note that, despite the low location of the floorboards between the frame, the tunnel for the driveshaft is an insignificant hump.

The Hornet probably has the distinction of having the flashiest steering wheel insignia in the business. As for the dash, instrument grouping is good, although reflections are numerous. Top of the dash, however, has a good non-reflecting surface for greater eye ease.

GENERAL

It's easy to say a great many good things about the Hudson Hornet. The most important of these are the performance and handling qualities, which have been adequately commented upon. But another vital characteristic is its ruggedness, a factor that has been conclusively demonstrated in dirt track competition. The road crew discovered this, too, for few cars emerge from the violent maneuvers the test drivers use in top condition. When the Hornet was returned to the factory, it still was in perfect tune.

Soup Up The Hudson Jet

By STEED EVANS

THE HUDSON Jet isn't going to break any sound barriers but in its two shakedown years it has made several more expensive cars look like World War I 'Jennies.'

SPEED AGE has been asked by many Jet owners if their cars can be hopped up. The answer is a qualified 'yes.'

At present there are no special speed parts being made for the Jet except those offered as catalogue equipment by Hudson. One can have hot parts made, at considerable expense, but it is unlikely that performance beyond that obtained with optional equipment will be gained.

The biggest obstacle to hopping up the Jet is its block. The sturdy alloy block will take a real beating, but, because of its bore placement there is little room for a deep bore job. Nos. 1, 2 and 3 bores are located to the fore of the center of the block, while the rear three are far back of center, leaving a wide water jacket between the two halves.

The sides of each three-inch bore are very close to the next in line.

Not enough metal can be ground from the bores for a significant change in displacement without chancing hot spots and warping. Some owners have tried and report that the maximum is about .050.

Such a cut would result in a gain of only eight cubic inches. This small gain, placed against the cost of oversize pistons and the bore job—$85—seems hardly worth it. A rise in displacement automatically raises the compression ratio, and the resultant knock on a Jet with 8.1x1 head, using regular gas, would simply mean detuning.

Another obstacle in hopping up the Jet is its 4.75 inch stroke. Of all popular make passenger cars, only the Chrysler Windsor Deluxe has one as long. A small bore and a long stroke do not make for continuous high speed, much less for drag racing. The engine will rev well enough at 5,000 rpm. but any more will strain its potential. Several East Coast drivers have had their clutch pressure plates disintegrate at 5,500 rpm.

But the Jet has its bright side, too. Hudson has always been interested in racing. There were Hudson cars whirling around the Indianapolis brickyard when today's racers were just learning to ride bicycles.

Today, Hudson makes a variety of optional speed kits for its big Hornets. Optional speed equipment is also available for the Jet.

Just by adding Twin H power—two single-throat Carter carburetors—and the aluminum head, a gain of about ten horsepower can be attained. Stock Jets are rated at 104 horsepower, with accessories mounted. Unlike some cars fitted with hot heads and dual jugs, the Jet performs smoothly throughout its speed range. There are definitely no flat spots in its acceleration curve with this optional equipment. Gas mileage is about the same as stock.

But the kicker is the availability of a high performance camshaft. This cam (Part No. SU 309777) is about the equal of a civilized full race grind. It lists at $32.50 plus tax and comes as part of what Hudson calls its 'severe usage' kit.

This cam can be installed without removing the head.

One mechanic, with nine years of tuning Hudsons (he prepared a Jet for Fonty Flock last year), said that his over-drive-equipped Jet with this cam, plus the optional head with .050 aluminum removed and Twin-H power, clocked 110 mph. The stock Jet with overdrive is rated at 93.5 mph.

Total outlay for the cam, installed and for shaving the head, plus gaskets, comes to $77. It is possible that a Jet with the special cam already installed could be ordered from the factory. Twin-H power and the 8.1x1 head, of course, are optional equipment.

Although it would be difficult, exhaust headers could be formed out of stock and this should improve the Jet's performance. The valves and valve train, as well as the distributor and coil in stock shape, are sufficient for the car's potential.

Stiffer shocks and a rear sway bar should be included for high speed running. The Jet unfortunately has lots of side roll. Even so, an optionally equipped Jet can take the new Ford hands down. If this engine had a blower, like the Kaiser, it should really go. ●

Hudson's Italia

Styled like no other car, Hudson's Italia has a sleek streamlined silhouette nearly 10 inches lower than standard Hudson models, which also incorporate the 'step-down' design. The Italia is built on a 105-inch wheelbase and powered by a 114 hp Jet engine. The chassis is designed to handle the powerful Hornet engine as well.

The body was designed and produced in Milan, Italy, by Carrozzeria Touring—one of the world's most famous builders of custom car bodies—in collaboration with Hudson engineers and designers. The broad, wrap-around 'panoramic windshield' has no posts to obstruct forward vision and functional airscoops, set into front fenders over the headlights, direct cooling air to the front brakes. Air intakes faired into the rear fenders cool the rear wheel brakes.

Driver and passenger can enter the low-silhouette Italia with ease since the doors are carried 14 inches into the roof. A triple-bank of chrome tubing faired vertically into each rear fender simulate jet stacks; actually, they hold the tail, signal and back-up lights. The front end is broad and low with the bonnet sloping below the fender line.

Inside, the revolution continues. Two individually adjustable 'Anatomical' seats—with reclined backs—are shaped to fit the contour of the shoulders, back, and hips. The passenger and driver can sit relaxed, without 'holding on,' even when the car corners at high speeds.

Upholstery is in fine textured red-and-white leather. A deep-pile, Italian-red rug covers the floor. An interior compartment behind the seats provides room for a vast amount of luggage. The dash is non-reflecting and finished in red.

Hudson officials say the Italia, with only a few modifications, could easily become a family car. However, they refuse to say when, or if, the Italia will go into production. ☆ ☆

A large American car which seats six comfortably, and has good performance

BY any standards the Hudson Super Wasp is one of the hotter saloon cars available on the Australian market.

It gave on an average of test runs a top speed of 89½ m.p.h. and its average standing quarter mile time was 25¼ sec.

For those not interested in performance it has looks, space, and that rugged luxury peculiar to American cars.

The Super Wasp is wholly assembled in Australia from imported parts, the main assembly plant being that of the N.S.W. distributors in Sydney.

Due to dollar restrictions some of the refinements on the U.S. made cars are missing. The car is available in America with vacuum assisted brakes and overdrive and optional extras are power steering and an automatic transmission. Also lacking on the Australian assembled cars, but standard on the American ones, are trafficators.

Apart from these the Super Wasp selling here is the same as its American counterpart.

It has the large hard working 4,293 c.c. side valve engine which develops 140 b.h.p. at 4,000 r.p.m. For Australian conditions a fairly low rear axle ratio of 4.09 to 1 has been chosen.

In the U.S. Hudsons are offered with a wide range of gear and axle ratios. The usual U.S. model gearing yields 26.6 m.p.h. per 1000 revs in overdrive.

The Super Wasp available here has the ratios, 11.78, 1st; 7.44, 2nd; and 4.09, top. The top gear yields 19½ m.p.g. per 1,000 r.p.m. with the 7.10-15 tyres which are standard here.

The Australian ratios are a sensible choice for all around road performance and the car can be driven fast and long without stressing the engine. However the lack of a suitable overdrive is felt on the score of economical operation.

If one thing is true of fast cruising cars like the Hudson it is that speed costs money. On a long undulating test stretch where it was cruised between 70 and 90 m.p.h. petrol consumption fell to 10 m.p.g

More moderate cruising speeds between 60 and 70 m.p.h. increased this figure to 14 m.p.g. Steady highway cruising with a limit of 50 m.p.h. brought it up to 18¾ m.p.g.

If driven gently the Hudson will turn in reasonable petrol consumption for most work. Its overall fuel consumption on the test was 16 m.p.g., while sane city driving gave 18 m.p.g.

The first thing that impresses about the Hudson is its size. By any comparison it is a big car. It is 16 ft 10½ in. long and its laden weight is 36¼ cwt. Those used to driving small cars, as the test drivers are, are straightaway dubious of getting a car the size of the Hudson through city traffic and worried whether they can find a city parking space long enough.

This feeling soon wears off, for the Hudson runs quietly and responsively.

There *is* a problem in parking but it disappears once you realise you have to find a large space, and what is so bad about size when a 6-person family can stretch out in comfort after throwing 23 cu. ft. of suitcases in the boot?

The steering is low geared with 5 1/3 turns from lock to lock on a 41½ ft. turning circle, but is adequate for most driving.

Fast driving demands fast steering ratios but the Hudson's size mentally blocks a driver throwing it into a corner. When it is pressed into a corner hard and fast the car surprises by its good handling.

There is some body roll and tyre howl, but there is a well-balanced understeer which makes a driver confident the tail will not break away.

Most American cars, in the hands of an experienced driver, can be cornered faster than is generally imagined.

With the Hudson this is very fast. The driver is warned he is overdoing it by the car digging on its suspension and working against it as though trying to throw off its load.

When the tail does break it does so quite gently and is easy to correct.

Proof of the Hudson's handling is the good showing several have made in N.S.W. reliability trials and the last two Redex Around-Australia Trials.

The car is very good on loose gravel roads and providing a little care is taken it will tackle corners well, but a corner must be deliberately lined up if the car is travelling at all fast. The Hudson is too long to slide on the narrow dirt roads we know in Australia.

The suspension handles rough stuff well. Corrugations mean little when the speed is above 30 to 35 m.p.h. The front end will not bottom unless the car is driven much too fast for the circumstances.

Dust sealing is very good on loose surfaces and there is little underbody noise for stones and gravel.

A further drive in the car after it had done 5,000 hard miles as a country demonstrator showed the truth of the Hudson's claim to be hard-wearing.

The shock absorbers were still tight, there were no suspension or body rattles.

Apart from the usual dust and dirt which collects around the door jambs, the car was clean inside.

Ignoring this, there was nothing to show that it had been in the country at all.

There is little effort to driving the car other than steering. All the controls are light and need only slight pressure. By far the best of these is the clutch.

The Hudson is one of the few cars, excepting those with automatic transmissions, that has kept to the old-fashioned wet-plate clutch. This has a completely shudder-free take-up at all times and stood up to hard acceleration runs without protest.

In city traffic particularly its easy operation made hard driving easy.

The Super Wasp is one of those cars that needs little effort to keep going. Top gear will deal with all normal road work, second gear only being needed for odd steep grades.

On level ground first, second, or top gear starts do not seem to worry the transmission apart from acceptable shudder when using top gear.

Working through the gears from rest on the level seems a waste of energy unless one is in a hurry. The gearbox is slow and a pause must be made when shifting from first to second and second to top.

The steering column lever which is mounted on the right adds to the slowness of the shift until one is accustomed to it.

Nevertheless high speeds are available on the gears if needed, 38 m.p.h. being possible in 1st, and 60 in 2nd. These are more than adequate for hedgehopping slow moving traffic on long grades.

The brakes are well up to the car's performance and fade only slightly under hard work. The drop in efficiency on a fast descent of the test hill was 5 per cent., which is rated as good for a car of this type.

The stopping distance of 31 ft. from 30 m.p.h. in neutral is also good, but the handbrake was not up to the footbrake in efficiency.

It is mounted under the right of the facia and is awkward to pull on. On the test car it was out of adjustment and took 206 ft. to stop

Above: The Hudson's engine is accessible, but it is a fair stretch over the high wide mudguards. Right: The front seat takes three comfortably, but will fit four at a pinch.

SPECIFICATION

MAKE:
Hudson Super Wasp, four-door, six-passenger saloon. Our test car from Ira L. & A. C. Berk Pty. Ltd., William Street, Sydney.

PRICE AND AVAILABILITY:
£2,185 (incl. Sales Tax); three months.

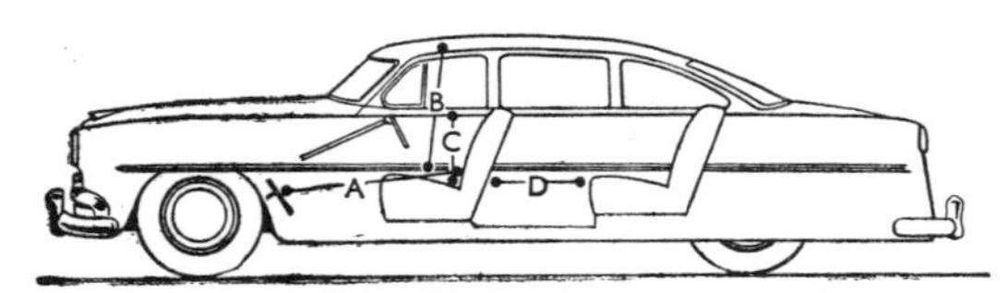

DIMENSIONS:
Front Seat: Width, 5' 0½"; squab height ("C"), 1' 9¾"; pedals to squab ("A"), 2' 11¾"/3' 1¾"; height over seat ("B"), 3' 0"; steering wheel to squab, 1' 2"/1' 4"; cushion depth, 1' 7½"; cushion from floor, 1' 0".
Rear Seat: Width, 5' 1½"; individual width, 1' 9½"; squab height, 2' 0"; cushion depth, 1' 7"; distance between seats ("D"), 1' 0¾"/1' 2¾"; height over seat, 3' 1½"; cushion from floor, 1' 0½".
Doors: Front, 2' 11½" x 3' 2¾" (avg.); rear, 2' 5½" x 3' 2¾" (avg.).
Boot: 4' 3¾" x 3' 3" x 1' 7¾" (avg. free space); spare wheel carried at side.
Overall: Wheelbase, 9' 11"; track: front 4' 10½"; rear, 4' 7½"; length, 16' 10½"; width, 6' 5-3/16"; height, 5' 0⅜"; ground clearance, 8⅛"; dry weight, 31½ cwt.

ENGINE:
6-cyl. s.v., 90.48 x 111.12 mm., capacity 4,293 c.c., comp. ratio 7 to 1, 140 bhp at 4,000 rpm, 2.35 bhp per sq. in. piston area, 214 lb./ft. torque at 1,600 rpm. Single Carter dual-down-draft carburettor with oil-bath air cleaner, by-pass oil filter. Sump 26 pts., radiator 71 pts., petrol tank 16-2/3 gals.

TRANSMISSION:
Single wet-plate clutch; three-speed gearbox with synchromesh on top, two ratios, operated by steering-column mounted shift lever; open propeller shaft; hypoid bevel final drive, ratio 4.09 to 1. Overall ratios: 4.09, 7.44, 11.78, reverse 14.32. Top gear mph: 19.5 at 1,000 rpm; 67 at 2,500 ft./min. piston speed.

CHASSIS:
Stressed skin unitary construction with separate front sub-frame.

SUSPENSION:
I.F.S. by coil and wishbones, rear by semi-elliptic leaf; telescopic shock absorbers.

BRAKES:
Foot-operated 4-wheel hydraulic with emergency mechanical linkage to rear wheels; hand-operated mechanical linkage to rear wheels from facia-mounted lever. Friction area, 140.35 sq. in.; ratios: 89½ sq. in. per unladen ton; 79½ sq. in. per laden ton.

ELECTRICAL EQUIPMENT:
6-volt ignition, 100 amp. hour battery, automatic courtesy light, dual horns, 35/45 watt headlamps.

STEERING:
Worm and roller; turning circle, 41½ ft.; 5-1/3 turns from lock to lock.

WHEELS AND TYRES:
Pressed steel discs. Tyre size 7.60-15. Recommended pressures: front 26 lb. per sq. in.; rear 24 lb.

PERFORMANCE

TOP SPEED:
Average of test runs	89½ mph
Fastest one way	90 mph
Highest recorded during test	95 mph

MAXIMUM SPEEDS ON GEARS:
1st, 38 mph; 2nd, 60 mph.

RECOMMENDED SHIFT POINTS:
1st, 10 mph; 2nd, 20 mph (see text).

MAXIMUM BHP ON TEST:
138 bhp at 4,110 rpm; equivalent top-gear speed, 80 mph.

MAXIMUM TORQUE ON TEST:
210 lb./ft. at 2,570 rpm; equivalent top-gear speed, 50 mph.

ACCELERATION:
0-80 mph through gears (see graph), 34.2 sec.
20-30 mph: 1st, 2.2 sec.; 2nd, 3.0 sec.; top 4.4 sec.
20-40 mph: 2nd, 5.8 sec.; top, 9.5 sec.
20-50 mph: 2nd, 9.5 sec.; top, 14.0 sec.
20-60 mph: 2nd, 13.6 sec.; top, 20.4 sec.
20-70 mph: top, 30.0 sec.
20-80 mph: top, 39.4 sec.

STANDING QUARTER MILE:
Average of test runs	21.25 sec.
Fastest one way	21.2 sec.

BEST HILL CLIMBING:
Top gear: 1 in 8.9 at constant 40 mph.
2nd: 1 in 5.9 at constant 30 mph.
1st: 1 in 4.2 at constant 18 mph.

BRAKING:
Footbrake at 30 mph in neutral, 31 ft.
Fade (see text), 5%.
Handbrake at 30 mph in neutral, 206 ft. (see text).

SPEEDO ERROR:
10 mph (indicated)-10 mph (actual); 20 mph-18¾ mph; 30 mph-27¾ mph; 40 mph-36¾ mph; 50 mph-47½ mph; 60 mph-57½ mph; 70 mph-69¾ mph; 80 mph-80½ mph.

TEST WEIGHT:
Driver, assistant, full tank, gear, 36¼ cwt.
Distribution: front, 20¼ cwt.; rear, 16 cwt.

FUEL CONSUMPTION:
Hard driving (see text), 10-14 mpg; highway cruising, 18¾ mpg.

The Hudson can seat eight at a pinch, has a vast boot and a big parcel-tray behind the rear seat. Overall finish and accessibility is good.

the car from 30 m.p.h. in neutral but was a satisfactory parking brake.

The Hudson's instruments are best described as following contemporary American practice. There is a large semi-circular speedometer calibrated from 0 to 120 m.p.h. but with final noughts dropped from the figures. Supplementary gauges are engine temperature and fuel which are respectively calibrated C-H and E-F. There are warning lights for oil pressure, ignition, and headlamps' high beam.

The speedometer was more accurate than it looked but the temperature and fuel gauges were too vaguely calibrated to be of much use.

Instrument lighting is mild but effective. All calibrations are clearly visible at night but are not glaring and do not reflect on the windscreen.

The headlamps are good for sustained speeds up to 70 m.p.h. and give a good spread for corners. They dip to the left and light the shoulder of the road and stop dazzle for oncoming drivers. When dipped they are satisfactory to speeds up to 50 m.p.h.

However, a supplementary driving beam would be helpful for long distance night driving.

The under-bonnet is very accessible and the maintenance should be easy. The only problem when decarbonising would be lifting the long cylinder head, or adjusting tappets.

Wheel changing is straight-forward at the front but on the enclosed rear wheels is more difficult. The covering spats must be removed and the wheel jockeyed between the brake drum and mudguard for fitting. Since the wheel and tyre are fairly bulky this could be trying.

The tool kit is adequate for all general maintenance but would have to be supplemented for serious work.

All told the Hudson Super Wasp is an impressive large car which has a degree of space and comfort unusual these days.

The interior is well appointed and has such sensibly-large items as man-sized ashtrays. Another uncommon feature today is a full width woven rug rail on the rear of the front seat squab.

Add to these points windows that wind down all round, excellent driver and passenger visibility, and quiet, draft free running and you have a car that represents sense at a reasonable price.

COMMENT FOR THE WOMAN DRIVER

by June Dally-Watkins

Although looking startingly large the Hudson Super Wasp is a car that makes driving deceptively easy.

It is very light to stear, the brakes and clutch untiring to use. And its engine pulls in top gear from speeds lower than a walking pace.

As a result it is a lot easier to get through traffic than some cars I have tried. But I think most women would be at a loss parking it.

It is very long to manoeuvre and the overhang at the rear is large enough to hit the car behind if care was not used.

This is a car in which you could take the family, inlaws, and close relations for a tour and still have room for luggage.

The front and rear seats are wide enough to seat four small people across and there is enough leg room for tall people. To stop two sliding on the vast rear seat there is a 1 ft 6½ in. wide arm rest.

Good points I noticed about the interior were arm rests on all doors, assist straps at the rear, a lockable glove box, and a very large parcel shelf behind the rear seat.

The interior is very well appointed. Upholstery is in genuine leather over Dunlopillo. Door and seat trim is also in leather which is two-toned to match the exterior.

The roof lining is in washable woven plastic which would make for easy cleaning as this part of the interior is a dust collector in most cars.

While getting the front seat out would be a two-woman job it would leave almost walking room for spring cleaning. The stepped-down floor must be vacuumed as it would not be possible to sweep the dirt out.

All told I thought it one of the better cars I have driven.

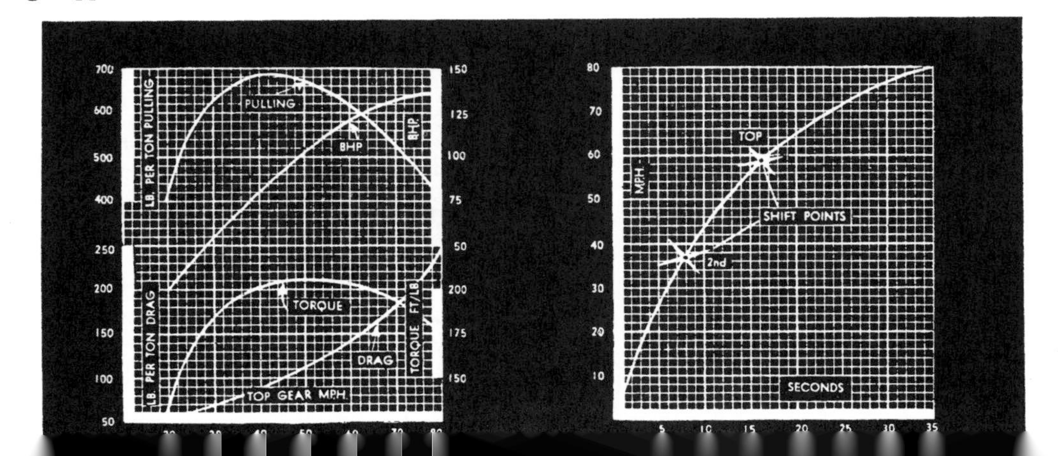

The Hudson's engine gave the imprèssive figures of 138 b.h.p. and 210 lb./ft. torque. Acceleration is particularly good, but fuel consumption is high when car is pushed.

For the first time since 1948, this old-line independent car has had a major face-lifting in styling

Front view of the 1955 Hudson Wasp shows the new, massive continental-styled grille with traditional Hudson crest in the center of upper frame. Hooded lights are recessed in fenders.

1955 Hudson

THE long-awaited Hudson styling change finally has appeared in the 1955 models. For the first time since 1948, this old line independent car has had a major face lifting. According to the Hudson Division of American Motors Corporation, the 1955 models 'represent the most sweeping change-over in the company's 46-year history.'

After the merger with Nash last year, there was much speculation as to whether Hudson styling would follow Nash in the European motif created by Pinin Farina and other Italian coach builders. The Hudson grille for 1955, among other body changes, answers the question affirmatively. Both the Hornet and Wasp feature the wraparound windshield and Hudson follows the Nash claim of a "completely distortion-free windshield.

The front fenders have been raised and the hood lowered, giving the driver a more definite idea of the whereabouts of his right front quarter.

Hudson buyers this year will have a choice of three engines in the Hornet series. A new V-8 of 208 hp., with optional twin Ultramatic transmission, is available. And, as usual, the reliable Championship six with or without twin carburetion can be had. The Wasp comes with Hudson's new 'Ti-Torque' L-head six cylinder power plant developing 110 hp. and can be equipped with dual carburetors.

All power accessories are available.

An exclusive feature of the completely restyled 1955 Hudson Hornets and Wasps is a 16-inch folding armrest for the convenience of passengers in rear seat.

The Hudson Wasp four-door sedan features many advanced engineering and style ideas—including wraparound windshield, air conditioning, deep coil suspension.

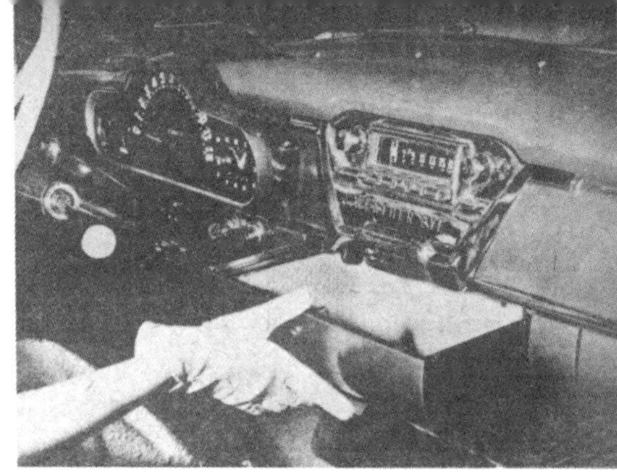

A large bin-type glove box in the center of instrument panel is a special feature of all 1955 Hudson Hornets and Wasps.

Completely restyled Hudson Hornet features sleek body lines, continental-styled grille, wraparound windshield, and an improved deep coil suspension. It reflects influence of Italian master, Farina.

Handsomely styled chrome letter 'H' serves as a convenient hand grip on the gas tank filler cap of 1955 Hudson models. The tail lights are mounted high on rear of fender.

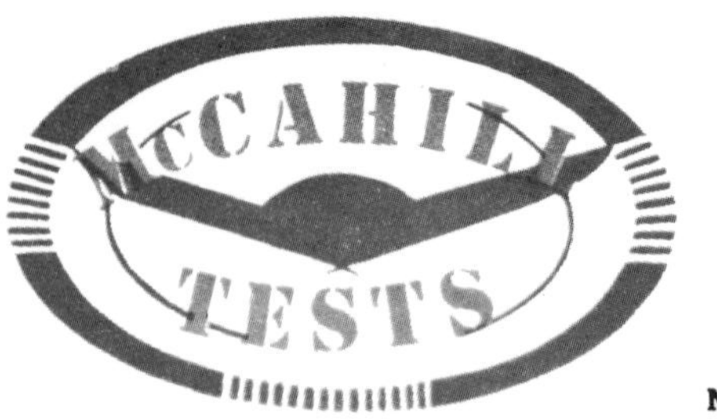

The Hudson Hornet

By Tom McCahill

A great, safe road car with brakes as fine as any he's tested, plus a gutty Packard engine "as smooth as silk shorts or a brandy Alexander," our Uncle Tom reports.

New Hudson, Tom finds, resembles models of old same as Harpo Marx does M. Monroe.

Fully loaded, 3,900-pound job stows enough gear to sink a Maine guide's freight canoe.

Tom was amazed when this buggy cornered with agility of the torsion-bar Packard.

THE Hudson automobile, a favorite with thousands long before Eddie Cantor was old enough to play juvenile parts, has just undergone the largest change in its long history. A year and some months ago, Nash merged with a declining Hudson outfit and came out as American Motors. Though this merger took place quite some time ago, the current '55 Hudsons are the first offerings under the American Motors banner, which is headed by George Romney.

Looking back briefly, right after World War I Hudsons were considered the smartest cars on the road, powered by what they called a Super Six. The cars of the 1919-1920 era, with box-like bodies and neat trunk racks on the roof, were topped with square little parking lights, miniatures of the carriage lights so common at the turn of the century. These Hudsons had all the smartness and chic of the Lincoln Continental which became the style leader two decades later. The Hudson boat-tail speedster was the Thunderbird of its day and, what might surprise some readers, at one time Hudsons sold for more dough than Cadillacs and were considered luxury products.

The years that followed saw Hudson giving birth to the once-very-popular Essex and, quite recently, the Jet. The post-World War II era saw Hudson snap itself out of a decline with the introduction of the Hornet, a big, husky, roadable six that dominated all stock-car racing for several years. Hudson was one of the first American cars to feature the step-down design and a 60-inch silhouette.

This new 1955 Hudson resembles Hudsons of old just about as closely as Harpo Marx resembles Marilyn Monroe. The present car is undoubtedly the first in the history of the automotive world that is now three inches higher off the ground than it used to be. American Motors claims its new over-all height is 62¼ inches "when loaded." Actually, in construction and body-wise, the new Hornet very closely resembles the Nash Ambassador with its combination frame-and-body construction built in one unit. This was undoubtedly the reason for eliminating the step-down feature and those extra three inches of height.

For power plants, these cars are still available with the old flat-top six that made Hudson champion of the stock car track. But the hot tamale of the clan is the Hornet V8 which is manufactured for American Motors by *Packard*. This V8, displacing 320 cubic inches, supposedly develops 208 horsepower at 4,200 rpm and has a twisting power of 300 foot-pounds of torque. Weight-wise, the Hudson's 3,900 and a fraction pounds, when fully loaded with enough equipment to sink a Maine guide's freight canoe, makes it on the lighter side for an engine of this displacement and power. Consequently, the '55 Hudson's performance is far from tugboatish. Zero-60 mph averages 12.4 seconds and 0-30 mph takes four seconds flat. This is not record-breaking performance for American sedans but, by the same token, it puts Hudson right in the big league and just an eye blink behind the fastest in a traffic light hassel. Top speed averages 106-109 mph, which is hardly loafing.

Parked at the curb these are good-looking cars, nicely appointed, and the rig I tested had a continental tire mount-

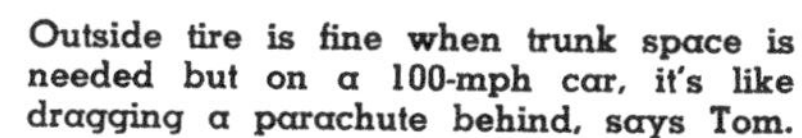

Outside tire is fine when trunk space is needed but on a 100-mph car, it's like dragging a parachute behind, says Tom.

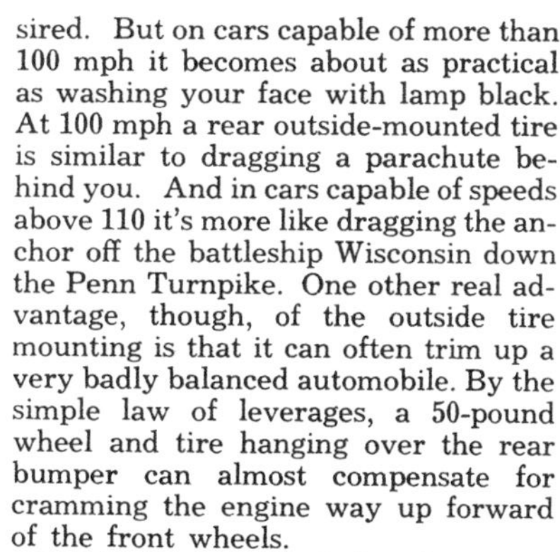

Hornet closely resembles Nash Ambassador with its combination frame-and-body construction built in one well-designed unit.

Here's Hudson's version of the wraparound windshield which the whole industry seems to have adopted. Note square side window.

goes broke. The other features are the all-weather air conditioning, which is the finest in the world, bar none—and most of Hudson's accessories, such as the excellently placed radio speakers, which give you a hi-fi effect in the cabin. You can buy a new Hudson with power lift windows, power steering, power brakes and all the luxury gook—or you can buy it bare, with standard stick or overdrive. In fact, you can buy the car the way you want it and there is no forced merchandising so common now among some offerings. These are great automobiles, worth every dollar they cost. If you're an old Hudson fan, you'll find the '55 Hudson your dish of tea. •

ing, which is just about as "continental" as the Irish stew flipped at you at Dinty Moore's. There is a tire well inside the trunk for those who don't go for this non-streamlining equipment. Back in the days when I was a cub scout in Larchmont and pearl gray spats were as popular as $1,000 bills, outside tire mounting was considered proper decor. I went for it on the Lincoln Continental and, years later, on so-called continental-equipped Fords, Ramblers, etc.

Outside tire mounting is truly functional when a lot of trunk room is desired. But on cars capable of more than 100 mph it becomes about as practical as washing your face with lamp black. At 100 mph a rear outside-mounted tire is similar to dragging a parachute behind you. And in cars capable of speeds above 110 it's more like dragging the anchor off the battleship Wisconsin down the Penn Turnpike. One other real advantage, though, of the outside tire mounting is that it can often trim up a very badly balanced automobile. By the simple law of leverages, a 50-pound wheel and tire hanging over the rear bumper can almost compensate for cramming the engine way up forward of the front wheels.

Inside the Hornet I found one big beef. The car I tested had an automatic transmission (Ultramatic) and the shift quadrant was about as easy to read as being elected Mayor of Milwalkee on a "No Beer" platform. And while I'm giving the designers of this barge such a happy time, let's consider the window roll-up handles. They are placed so far forward and down on the door that unless you are a graduate of a Parisian school for pickpockets, you might find them very hard to reach. It will be good for your shape, though, and should result in a belly as flat as Khrushchev's head. Now that we've got these few little observations out of the way, let's take the car for a ride.

Hudson became famous in racing through superb roadability. The 1955 Hudsons have lost none of this but have developed an odd characteristic rarely found in top-flight road cars. In hard bends and corners, the front fenders drop down and apparently plow like a Flexible Flyer going into a soft snow bank. Oddly enough, the wheels stay true and the car gets around the tightest bends just as well as it ever did. This new Hudson had the exact same breakaway factor that I found in the new Packard with torsion-bar suspension. As I tested both these cars on the same course under the same conditions, I was quite amazed, after the fender dips, that both cars held the rear steady before sliding out up to 52 mph on my hardtop corner.

Scientifically, this means nothing to you or anyone else, as all corners vary, but I use the same corners on many of my tests for comparison. I keep a record of how each car handles on each corner and where the breakaway point occurs. On my dirt corner I also found this car got around like the Packard, which was excellent, except that I could actually hold it a little further through before going into a slide. These Hudsons are great, safe road cars and the brakes are as fine as any I've ever tested on an American car. The Packard engine in this job, aside from being gutty, is as smooth as silk shorts or a brandy Alexander.

Some of the features that Hudson has borrowed from Nash—and which are by far the best in the industry—are: number one, the reclining seats which can be made into full-length beds, either single or double, which means one guy can drive while the other dreams of making

SPECIFICATIONS

MODEL TESTED:
1955 Hudson Hornet four-door sedan

ENGINE:
8 cylinder, V-type; bore 3 13/16 inches, stroke 3½ inches; maximum torque 300 foot pounds @ 2000-2600 rpm; brake horsepower 208 @ 4200 rpm; compression ratio 7.8 to 1

DIMENSIONS:
Wheelbase 121¼ inches; overall length 209¼ inches; tread 59½ inches front, 60½ rear; width 78 inches; height 62¼ inches; weight 3,900 pounds as tested; standard tire size 7.10x15; gas tank 20 gals

PERFORMANCE:
0 to 30 mph, 4.0 seconds
0 to 50 mph, 8.1 seconds
0 to 60 mph, 12.4 seconds
0 to 70 mph, 17.3 seconds
Top speed, 106-109 mph

SPEEDOMETER ERROR:
At 60 mph on speedometer, actual speed 56.3 mph

PRICE:
$3,675 delivered in New York

America's top (and only) trunk tester, Jim McMichael, squats his bulk in the storage space. His qualified opinion was, "Roomy."

Nicely appointed inside, Hudson's reclining seats convert to double beds and Uncle T says air conditioning is world's finest.

NEW HUDSON GRILLE

NEW NASH GRILLE

NASH AMBASSADOR

Hudson and Nash Get V-8 Engines

NEW body styling—the first major change since 1948—and new engine choices are the main 1955 headlines from Hudson. Optional on the Hornet, the top car in the line, is a big OHV V-8 engine rated at upwards of 205 horsepower. Smaller L-head sixes rated at 170, 160, 114, and 104 hp. are also available, as are manual, overdrive, or Hydra-Matic transmissions and air conditioning. Other features include reclining seats, twin beds, a new cowl-width air intake, visored headlights, and wrapped bumpers.

For Nash, a similar V-8—which closely resembles the smaller of the two new Packard engines (see page 154) and which is reportedly sold to American Motors by Packard—powers the Ambassador, the top car in the 1955 line. Like Hudson's, it is a 320-cubic-inch engine of modern design. Sixes of approximately 140, 130, 110, and 100 horsepower are also available, the first two on the Ambassador and the others on the Statesman. Handiest recognition feature: Headlights have been plucked off the fenders and tucked in the grille.

All new this year is the front-end styling of the Hudson. Shown here is the 1956 Hornet V8 four-door sedan

'56 HUDSON

Hudson has a new grille that makes it instantly recognizable as it comes down the highway. The grille, a broad mesh design, features a bold "V" in its center, a styling motif that is repeated in Hudson's side treatment as well as in its interiors.

A narrow band of gold color that runs horizontally along the rear fender makes three-tone cars actually four-tone and two-tone cars three-tone.

All 1956 Hudsons use the 12-volt electrical system. Three basic engines are available and, in addition, two optional six-cylinder engines. Horsepower is up in all of them.

The Hornet V8 engine develops 220 horsepower with a compression ratio of 9.55:1 and a 352-cubic-inch displacement.

The Hornet Six develops 165 horsepower with a 7.5:1 compression ratio. An optional twin-carburetor kit increases the power to 175 horsepower.

The Wasp Six develops 120 horsepower and can be obtained with twin carburetion which boosts output to 130 horsepower.

A redesigned dashboard features a horizontal thermometer-type speedometer. A crash pad of foam plastic covers the top of the dash. Reclining seats that make up into beds are optional.

Continental tire adds 10 inches to the length of the hardtop. Narrow panel on rear fender is gold colored

CLOSEUP OF THE

HUDSON ITALIA

Now on the market, this car dreamed up in Detroit is a forecast of what's ahead

HUDSON'S ITALIA is not a true production vehicle yet, but many of its features are a good bet for assembly line cars of any manufacturer in 1956 and 1957. Only a few have been handbuilt in Italy, by Touring, and put on sale in Hudson dealerships. One of these was made available to MOTOR *Life* by Fullerton-Loadvine, of Santa Monica. Chassis and engine (114 hp) are Hudson Jet and the handling characteristics, therefore, compare favorably, if not surpassing, the pre-1955 Hudsons. Some hard corners taken at speed were easy and sure. Acceleration, while not clocked, obviously ranks with the best of sports-type cars in the low-powered class. Steering is quick by American standards. •

Windshield is contemporary, as is ventilation intake. Visibility is exceptional, however, since car is unusually small, gives feeling of good control at speed.

Door cut into top is one feature likely to be adopted in production designs. It makes sliding into bucket seats easier, but curved door top gets in the way.

Italia is a two-seater so rear of passenger compartment is set up for luggage. Interior, including dash (see front cover), is extremely simple but neatly arranged.

Triple-chrome tubes look like radical exhaust system, but actually are stop, backup and turn lights. Exhaust is conventional single pipe under rear bumper.

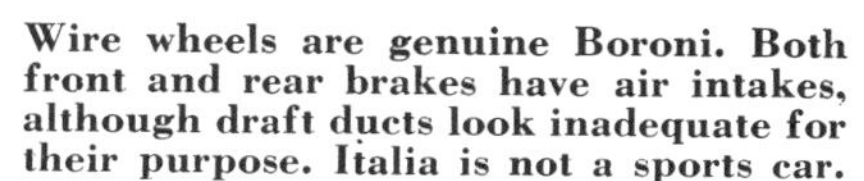

Wire wheels are genuine Boroni. Both front and rear brakes have air intakes, although draft ducts look inadequate for their purpose. Italia is not a sports car.

Rear window is recessed, an item expected from Detroit. Italia body is aluminum, which keeps weight down but transmits engine, road noise without undercoating.

American Motors' private test grounds provide surfaces and conditions that are comparable to the roughest a driver might find anywhere in the country.

HUDSON HORNET SIX

This year's models look less like Nash, and are greatly improved in performance, handling and roadability

LAST year was the first for Hudson as a division of American Motors; it was also the first in several years that saw Hudson down at the bottom in stock car racing. But 1955 was also the first year, for several recent ones, for Hudson to show a sales increase. Largely regionally, and mostly in the West and Midwest, Hudson dealers perked up and got something like 45 percent more business than in 1954.

Hudson management goofed by sticking too long to an outmoded style. When the merger with Nash came, management decided that the old "stepdown" floors would have to go. Instead, Hudson was given the same basic body as Nash, and this, more than anything else, probably led to its almost immediate decline as a favorite in stock car racing. Not that *unitized* construction which welds the frame and body together in one rigid structure isn't good (it is about twice as resistant to torsional stress as conventional chassis plus body is), but the resulting slightly higher center of gravity and the coil-sprung rear suspension just didn't enable the Hudson to corner on tough tracks as well as previously. Sales, too, probably suffered because the Hudson looked too much like the Nash, especially from the rear. Of course the Hudson's front wheels are exposed and front tread is considerably wider, making for better front end stability and shorter turning diameter than the Nash. In many ways the 1956 Hudson is much improved, and in our opinion this car can succeed if the public appraises its virtues with intelligence and considers the fact that this car offers more miles without body rattles than most of the larger production models.

For 1956 Hudson offers new visage: a mesh type grille, die cast and not stamped, has been installed that seems to increase the car's overall width, the headlights are topped off with decorative pseudo air scoops and the parking lights have been combined with new chrome trim to give that more massive appearance that Americans demand. The hood has been shortened slightly to accommodate the new V-shaped grille, but not lowered; yet visibility forward is better than average and the driver can easily see both front fenders. An airscoop extends across the entire hood in front of the windshield to feed air to the excellent Nash-developed Weather-Eye heating and ventilating system, or to the optional air conditioning unit. The latter is about the most compact in the industry, and it should be noted, one of the best at the lowest price.

On the Custom series, the more plush and stylish models which are available with the famed Hornet "Six" engine, three-tone color schemes are offered. New chrome along the belt line features a V-shaped dip toward the body rear; in this V-dip the symbol denoting the six-cylinder model is located. On these Custom jobs a gold-colored panel shaped like a parallelogram is inserted attractively between the two horizontal chrome strips giving the effect of a dash of a fourth color on the basic three-tone models. Super models, the same but with less ornamentation inside and out, do not feature three-tone schemes, but do offer 15 separate two-tone stylings. Either the Super or Custom models are available in 14 solid colors.

Restyling has continued to the rear of the car where new tail-lights, elliptically shaped, have replaced the previous units which looked too much like those on the Nash; and new rear deck chrome has given the Hudson a distinctive appearance. A continental type spare tire mount is optional but has two distinct disadvantages: the overall length is increased 10 inches over that given on our accompanying specifications, and the upright spare tire does not lean outward sufficiently far to make loading the luggage compartment anything but a laborious process.

Hudson has not gone to the exhaust-through-the-bumper vogue; this has probably been wise inasmuch as Chrysler, one of the Big Three, is getting away from this fad. Nor are dual exhausts available as factory options on the Hudson Hornet Six. The rear view of a Hudson shows a cleanness of line and a simplicity that has become something of a rarity on domestic automobiles. One improvement would certainly be to enlarge the deck lid that the opening would be flush with the level of the bumper.

Our test Hornet Six was a four-door sedan that had seen several thousand miles of hard proving ground work and which was being used as a press demonstration car. It might be well to mention here that many of the cars tested in this volume are early production models. Where possible we select our test cars at random, but this is not always possible, there being usually only one or two of each make ready for such activity at such early dates.

The back areas of American Motors test grounds resemble the worst possible type of roads; other areas include torture tests and washboard roads that no car should have to endure. The Hornet handled beyond our expectations: we learned that new valving in the rear shocks and different spring rates were engineered into the 1956 model. Rear end sway and body lean on hard corners has been largely eliminated.

We weren't able to drive the car at night, but sitting in dark shade with the instrument lights on indicated that there will be very little glare in the windshield. A padded top on the dashboard is standard on all Hudsons at no extra cost, and the dull finish of the padding helps to relieve eye strain. The sun visors are dark blue transparent plastic; between them is a knob that permits lowering the antenna. The instrument layout is completely new for 1956; we like the red-line type speedometer and the clock with the sweep second hand —cannot buy, however, the almost universal warning lights for oil pressure and ammeter (Hudson started this craze way back in the early thirties). The glove bin is large and well centered; optional radios have a speaker at each side, and all controls are easily reached. The Hydramatic indicator quadrant is a bit hard to see unless one is of greater than average height, and the mirror is much too small.

All in all, though, the Hornet Six is a good performer, should be economical, and the increased power, the optional dual carburetors, and the new hydraulic valve lifters assure the Six's staying in the 100 m.p.h. league. Ride is soft, roadability much better.

Handling qualities of the Hornet Six have been greatly improved this year; new rear shock absorber valving and changed spring rates help enormously.

From the rear, the new Hudson doesn't bear as close a resemblance to Nash as it did in '55. The so-called continental spare tire mounting, however, makes loading the trunk excessively awkward.

PERFORMANCE:

ACCELERATION, corrected speeds (from standing start):
Zero to 30 mph: 4.6 seconds (speedometer read 33)
Zero to 45 mph: 9.4 seconds (speedometer read 49)
Zero to 60 mph: 14.8 seconds (speedometer read 66)
HIGHWAY ACCELERATION (with stepdown) 50 to 80 mph: 14.3 sec.
MAXIMUM SPEED: 98-100 mph.
FORWARD VISIBILITY over hood: Approx. 18 feet

HUDSON: Other models: *Hornet V-8:* Dimensions same as Hornet 6. Engine: V-8, 352 cu. in., 220 h.p. @ 4600 rpm., 320 ft. lbs. torque @ 2200-2500 rpm. *Wasp:* Wheelbase 114-¼ in., Overall length 202-¼ in. Engine: 6-Cyl., 202 cu. in., 120 h.p. @ 4000 rpm., 158 ft. lbs. torque @ 1400 rpm. Optional 130 h.p. @ 4000 rpm, 168 ft. lbs. torque @ 1800 rpm.

The air scoops seen above the headlights are only false fronts, having no purpose but decoration; but there's a real one in front of the windshield.

SPECIFICATIONS

(all in inches unless noted otherwise)

HUDSON Hornet "Six"

CHASSIS & BODY:

Wheelbase	121¼
Tread—(front, rear)	59½, 60½
Length overall	209¼
Width overall	78
Height overall	62¼
Ground clearance	6-13/16
Turning circle diameter	42 feet 8 inches
Steering wheel stop-to-stop	3½ turns with power
Tire size	7.10 x 15
Weight (shipping)	3530 lbs.
Overhang—(front, rear)	34-15/16; 53-1/16
Brake lining area	192 sq. in.
Weight to brake area ratio	18.4 lbs. per sq. in.
Weight to power ratio	20.1 lbs. per BHP

ENGINE & DRIVE TRAIN:

Cylinders, block, valves	6, In-Line, Overhead
Bore and stroke	3-13/16 x 4½
Displacement	308 cu. in.
Compression ratio	7.5
Brake horsepower (maximum)	175 @ 4000 RPM
Torque	278 @ 1800-2800 RPM
Carburetor	Dual, downdraft
Fuel pump	Mechanical
Fuel tank capacity	20 gallons
Exhaust system	Single
Crankcase capacity	7 qts. with filter
Drive shaft type	Torque tube
Rear axle ratio (and available transmissions)	3-speed Manual: 4.1; Overdrive : 4.4; Automatic : 3.15
Cooling system capacity	19½ qts. with heater

INTERIOR DIMENSIONS:

Shoulder room	61½ front; 61¼ rear
Headroom	37½ front; 36 rear
Legroom	42½ front; 40⅛ rear

AN AUTO AGE STAFF REPORT

Nash Ambassador, above, and the new Hudson Hornet both use the same basic body shell, but front-end styling is vastly different.

HORNET and AMBASSADOR

AMERICAN MOTORS' BIG BROTHERS

No longer operating as independent manufacturers, Nash, Hudson are pulling each other out of danger.

Both cars have extremely wide and comfortable front seats, complete with reclining backs. This is the dashboard of the Nash.

WHEN THE HUDSON-NASH merger was announced and American Motors was formed, many skeptics considered the move to be the last futile gasp of two dying "independents." Hudson owners in particular, a wildly partisan group, bemoaned the passing of their beloved "step-down" stock-car champions and thought of the Nash in terms of low speed, great comfort, but worse-than-average handling.

But Hudson owners for 1955 were pleasantly surprised. They had been picturing the old Nash; the new one, while retaining the same basic body shell, had some new features. They were good new features, too, and ones that Hudson was to benefit from.

The 1955 Nash Ambassador and

On left is the 208-hp Packard-built engine that powers the Hornet and Ambassador. Suspension features high-mounted coil springs.

Hudson Hornet V-8s are admittedly very similar cars. They share a common body shell, chassis, engine and transmission and are available with exactly the same optional accessories. Their greatest variation shows up in front-end styling with the possibly-undecided buyer getting a choice of covered or uncovered wheels, headlight placement, grille shape and fender contours. Of the two, the Nash is the more unusual, with the headlights taken from their usual position in the fenders and set into the oval, Nash-Healey-type grille. Actually, we prefer the cleaner, more orthodox Hudson fender and light treatment, largely because of the fact that wheels are uncovered to a greater degree, allowing for a slightly smaller turning circle, but that is a matter of taste. The real advantage to this "marriage" is that both cars profit by it. Let us see how, point by point.

Let us take, for instance, the matter of comfort. For years Nash has been famous for its reclining seats and highly-efficient but inexpensive air-conditioning system. These are, of course, available on the 1955 car but can also be had now on the Hudson. The seats, in fact, are standard equipment and an air mattress with built-in pillows can be had for just a few dollars more. No amount of words can adequately describe just how wonderful this feature can be on a long trip. Driving position in both cars is quite easy to get used to and visibility through the widest wrap-around windshields in the industry is excellent. Leg room for rear-seat passengers is generous, even with the front seat pushed all the way back.

Next we consider roadability. Now this has never been a particularly strong point for Nash, while it definitely *has* been for Hudson. The new American Motors' cars are really a compromise. As far as cornering is concerned, the Nash is better than it has ever been (though not yet as good as the Rambler) and the Hudson is almost, but not quite, on a par with last year's car. Both of our test cars were equipped with power steering that went four turns from lock to lock, and it felt very good, with really surprising correction control. There is rather a lot of lean on tight corners, but the cars aren't at all skittish and traction is, if anything, above average. As to riding qualities, these cars soak up bumps with the best of them and while they do dip and bounce some when going over bad bumps or ruts, there is no loss of control, even at relatively high speeds. You'd have to go far to find a car at any price that can give a more luxurious ride. It might be said that Hudson has adopted Nash's ride and Nash has adopted Hudson's handling qualities. At any rate, they are both fine from these angles.

When it comes to the engine, transmission and general perform-

Grille front and headlight placement on the Hudson, left, are more orthodox than the Healey-inspired Nash treatment shown at right.

Nash looks neat from the rear and that huge rear window makes for wonderful visibility. Those two posts don't get in the way at all.

ance, however, it is not Hudson or Nash, but Packard who can take a bow. For the Ambassador and Hornet are both equipped with Packard engines and Twin-Ultramatic transmissions and while performance in either case will never win a Grand Prix, both of these buggies are a whole lot faster than they were last year.

Actual performance will vary according to what equipment the particular car is carrying. Not only do some of the optional luxury gadgets rob you of horsepower—they add weight. Our acceleration and top speed figures for the two cars vary, therefore, but under identical circumstances the Nash and Hudson should perform alike.

In our acceleration tests, we found that the cars were at their best when the Ultramatic transmissions were started in the lower drive position, which takes advantage of gears in conjunction with the torque converter. Running the cars up to about 55 mph in this range before shifting into the higher drive position gave really brisk acceleration, as the British would say. Taking an average for both cars, we got times of 4.4 seconds to 30 mph, 6.7 seconds to 40, and 12.3 seconds to 60 mph. Top speed runs gave figures varying from 102 to 107 mph. For an actual breakdown of our times for both cars, take a look at the chart in the next column.

We have long praised the Hudson braking system, with its extra mechanical brakes standing by to save your life in the event that the hydraulic gear should fail. This year, the Hudson and Nash have what are possibly the best brakes on any American car. Not only are the stopping distances below average, but the cars stop in a straight line from any speed up to and even past 60 mph, without any trace of break-away. This is a very important point, and a tribute to the chassis design, as well, for it is not enough to stop the wheels. You must stop the car, too, and without losing control. Too many modern cars have a nasty habit of spinning like tops when the brakes are slammed on in a panic stop from over 40 mph. This is definitely not the case with either the Nash or the Hudson for '55, and we salute them for it.

As for servicing, especially of the engines, we would definitely not recommend that anyone try to monkey with either of these cars unless he is a first-class mechanic. Not that there is anything so radical about them—it's just that the big V-8 power plants fit into those engine compartments as if they were poured in in liquid form, and some of the parts are almost impossible to get at without removing pounds of hardware. You would certainly have to take the engine out of the car to do a major overhaul, but, on the other hand, the Nash and Hudson V-8s should go many thousands of miles without needing such drastic repairs.

Taking a look at some of the other common features, we found that general finish was quite good, with no uneven seams or bad spots in the paint. These cars are built to last, if that happens to be important to you, and if it isn't, at least it's nice to know that they are rugged, especially if you should happen to bang into anything. They are almost entirely rattle-free and are apt to stay that way, with their unit construction. Price-wise, the Hudson is just a little higher, the basic price of a custom four-door sedan being $3,015 as opposed to $2,965 for the Nash. Either one, weighed against the present competition, is a fine buy. ●

Hudson eliminates the rear window posts entirely and cuts the fender lines higher, but has the same trunk-obstructing tire mounting.

SPECIFICATIONS*

ENGINE: V-8, overhead valves; bore, 3.81 in.; stroke, 3.50 in.; total displacement, 320 cu. in.; developed hp, 208 at 4,200 rpm; torque, 300 ft./lbs. at 2,000-2,600 rpm; compression ratio, 8.25 to one.

TRANSMISSION: standard is Ultramatic, 4-element torque converter with planetary gears; power steering unit available for additional $140.

REAR AXLE RATIOS: 3.54 to one.

DIMENSIONS: wheelbase, 121.25 in.; front tread, 59.5 in.; rear tread, 60.5 in.; width, 78 in.; height, 62.25 in.; overall length, 208.25 in.; turning diameter, 44½ ft. for Nash, 42 ft. 8 in. for Hudson; turns lock to lock, 4.3 (4 with power steering) for Nash, 4.6 (4.3 with power steering) for Hudson; weight, 4,200 lbs. for Nash (includes air-conditioning), 3,900 lbs. for Hudson; tires, 7.10 x 15 (tubeless).

PERFORMANCE

ACCELERATION	Nash	Hudson
0-30 mph	4.6 seconds	4.2 seconds
0-40 mph	6.9 seconds	6.4 seconds
0-60 mph	12.5 seconds	12.1 seconds
30-50 mph	6.9 seconds	6.7 seconds
TOP SPEED:	102-103 mph	106-107 mph

* Differences between cars are indicated where they occur.

NASH AND HUDSON

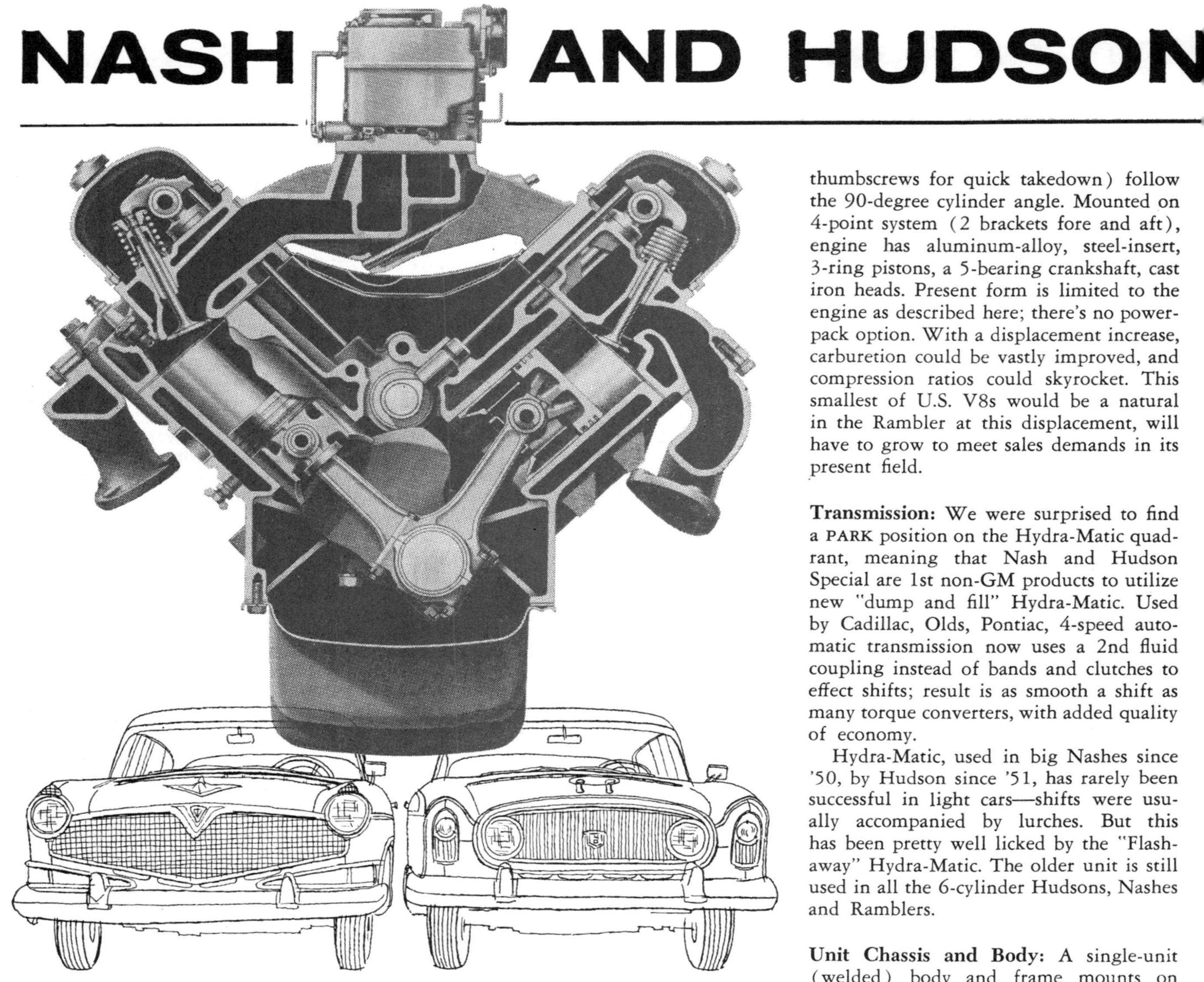

Four-door sedans with new Hydra-Matic: Nash with power brakes and steering, Hudson without

CUSTOMERS—potential or regular, high-income buyers or moderate wage-earners—are becoming more and more aware of what the magical term V8 means when compared to an old-type 6-cylinder engine. And American Motors has been more aware of this than any of its customers. The logical conclusion? Do something about it—fast! Long known for economy, durability and good riding qualities, these models needed a shot in the arm. A-M took a brand-new V8 engine, the name and appearance from more expensive, more powerful models (Hornet and Ambassador), added a smoother transmission, and tacked on the word "Special."

Engine: Corporation's own V8 with modest 250-cubic-inch displacement, about-average 8.0 to 1 compression ratio; 240 pounds-feet torque is on the low side (by today's standards) but has a good curve, flattening out at 2000-3000 rpm. Carburetion is 2-barrel Carter, exhaust system is via 3-branch outlet with a crossover running forward of the oil pan; the single tailpipe takes off from the right side of the engine. Engine looks quite conventional; its short, fairly wide rocker arm covers (topped by a pair of knurled, man-sized thumbscrews for quick takedown) follow the 90-degree cylinder angle. Mounted on 4-point system (2 brackets fore and aft), engine has aluminum-alloy, steel-insert, 3-ring pistons, a 5-bearing crankshaft, cast iron heads. Present form is limited to the engine as described here; there's no power-pack option. With a displacement increase, carburetion could be vastly improved, and compression ratios could skyrocket. This smallest of U.S. V8s would be a natural in the Rambler at this displacement, will have to grow to meet sales demands in its present field.

Transmission: We were surprised to find a PARK position on the Hydra-Matic quadrant, meaning that Nash and Hudson Special are 1st non-GM products to utilize new "dump and fill" Hydra-Matic. Used by Cadillac, Olds, Pontiac, 4-speed automatic transmission now uses a 2nd fluid coupling instead of bands and clutches to effect shifts; result is as smooth a shift as many torque converters, with added quality of economy.

Hydra-Matic, used in big Nashes since '50, by Hudson since '51, has rarely been successful in light cars—shifts were usually accompanied by lurches. But this has been pretty well licked by the "Flash-away" Hydra-Matic. The older unit is still used in all the 6-cylinder Hudsons, Nashes and Ramblers.

Unit Chassis and Body: A single-unit (welded) body and frame mounts on 4-coil springs backed up by direct-acting Monroe shock absorbers; drive is via torque tube. These cars, like the Statesman and Wasp, have a 114¼-inch wheelbase; only dimensional difference between the Specials is in Hudson's wider front tread—at merger time, Hudson brought along its own steering and kingpin setup, retains it on Hornets and Wasps. Mechanical reserve brake comes only without power boost.

Large for their class, the Specials (without the continental tire mount, which adds 10 inches to overall length) fall in the Plymouth (205-inch), Super 88 (203-inch) class in length. With the tire, the cars need the garage space of a Dodge or an Olds 98 (both require a long 17 feet, 8 inches).

Body differences between the Nash and Hudson are in the grille, tail lights, rear

AN MT RESEARCH REPORT BY JIM LODGE

SPECIAL V8 ROAD TEST

Body sway is weakest link in both cars' roadability and ride (Nash shown)

fenders, hood, rear top-quarter panel and rear window wraparound. Inside and out, trim follows that of the larger Ambassador and Hornet.

DRIVERS' COMMENTS

Driving position: High steering wheel has thick rim which, at 1st, makes the wheel seem smaller than it really is. (It's actually as large as an 18-inch Cadillac wheel, larger than that on a Plymouth or Ford.) Legroom isn't as great as in a Chevrolet or a Ford, but it's not skimpy. Nash-Hudson headroom, however, is unmatched by a Crown Imperial or any other high-, medium-, or low-priced car. Front seat is relatively high, tho not enough to make steering wheel bump your leg.

Ease of Handling: After driving a standard-steering Hudson and the power-steering Nash, we vote for power steering in these cars. Without it, steering is exceptionally stiff when parking, wheel requires a healthy tug to start front wheels around a turn at traffic speeds. On the highway, however, steering is accurate, with no feeling of looseness; directional control is good. Hudson requires about 1/3 more wheel turns (non-power steering) than the Nash when neither one has power steering; with it, the Nash is still slightly faster-steering by the book, but is not so on the road.

With power steering on either car, there's an amazing change to quick movement even tho reduction in lock-to-lock wheel movement is only about a half-turn. (By comparison, many Chrysler products drop from 5 or 5½ turns down to 3½.) The immediate relief on the Specials is most gratifying. A-M's Monroe power steering is a steady-steering setup, not too touchy, with a fair degree of road sense transmitted to the driver.

With its narrower tread, Nash Special should turn in less space than Hudson, but former's wheel stops (a hangover from shrouded front wheel days) dictate a 43¾-foot turning diameter (wall to wall), giving the Hudson the edge with a 39¼-foot circle. Fat, raised steering wheel hubs can strike your forearms when you're turning the wheel, should disappear in '57 models, when the majority of wheels will follow Ford's dished theme.

Nash Hudson

Vision: Coincidental with ease of handling is the bonus of being able to see the road close to the front of the car, or to zero in on the front-right fender when squeezing thru a crowded parking lot. Both cars have short front ends (with Hudson's seemingly shorter because of stepped hood styling), both have benefit of high fender lines; Hudson's simulated fender-top airscoops act as an added guide to car width. Rearward vision is good, and lack of tail fins or uplifted rear fenders goes unnoticed.

A-M follows the small rear-view mirror policy, preferring to sacrifice wide-angle view to the rear for a mirror that doesn't constitute a forward blind spot. Both cars were equipped **Continued on next page**

Nash

NASH AND HUDSON

with instrument panels padded around the top, effectively shrouded instrument housings to eliminate daytime glare and nighttime lighting reflections. "Hot spots" came from heavily chromed trim near center of dashboards, chromed steering wheel hubs and crossbars.

Instruments and controls: Nash instruments are placed toward the center of the panel, the Hudson's are grouped directly in front of the driver. Nash utilizes a conventional needle-type speedometer, has small-sized indicator lights instead of oil pressure gauge and ammeter. Thermometer-type horizontal speedometer is used in Hudson; it's as useful as a pointer type, but moved back and forth with evenness of road surface. Hudson instruments are easier to read, being white on black; indicator lights are huge, jeweled jobs that eliminate guesswork. Fuel and temperature gauges aren't as well placed as the high, line-of-sight gauges on the Ambassador Special. Hydra-Matic quadrants are placed farther down the steering column than on most cars so equipped. Altho Nash, with power brakes, had larger pedal than Hudson, neither car had brake wide enough to be used naturally with the left foot. Hudson pedal seemed ridiculously small by current wide-pedal standards.

HOW THE CARS PERFORM

Acceleration: Need for increased displacement quite evident here from a competitive standpoint. Smoothness of Hydra-Matic was appreciated, but this transmission with a 3.54 (or particularly the optional 3.07) axle isn't conducive to high performance with a low-torque engine. Snappier performance can be expected from overdrive-equipped Specials using the 4.55 to 1 axle supplied. With this setup, 0-60 mph times should be shaved to the 12½-13 second range; quarter-mile times should drop to about 18½-19 seconds, but don't expect much increase in speed thru the quarter. Low-speed acceleration thru 1st and 2nd gears should be excellent.

Times shown in the table were made in DRIVE range (using all 4 gears); transmission couldn't be outsmarted by using LOW or D-3 range—it shifted thru all gears no matter what range we held it in, a safeguard against over-revving the engine.

Braking: The Specials behaved oddly compared to other cars. They showed absolutely no swerve or uneven stops until the brakes faded completely on the 6th hard stop. "The Big Surprise" came after the 6th stop, when the brakes began to recover quickly. Within a minute after this stop another check was made, with full brake recovery. Pedal action remained hard for some time, and by the 11th and 12th stops there was slight pull to the left (which eventually disappeared), but no more fade.

Roadability: Manual steering gave uncomfortable feeling if the car hit soft road shoulder, required plenty of muscle to crank the wheel out of a rut. Generally well-behaved (no strain in streetcar tracks, no loss of traction bounding over railroad tracks at any speed) under all conditions, the Specials were susceptible to wind wander, lost normally good directional stability

Hudson

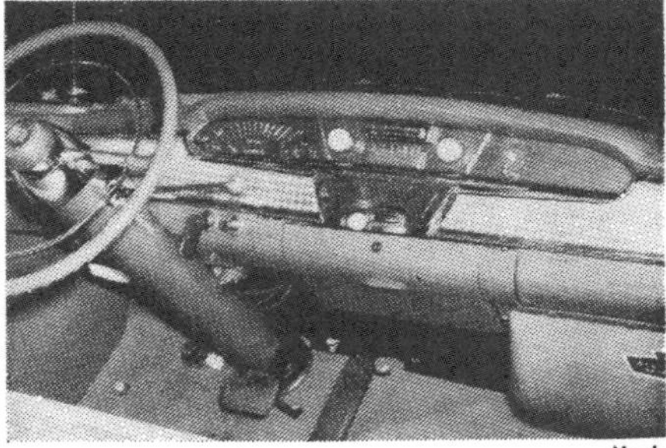

Nash

when cutting thru gusts of wind. Coming out from behind a passing truck at high speed with a stiff crosswind blowing, car would move out of its lane, with body movement removed from car direction.

Body lean was disconcerting in hard cornering maneuvers, but cars stuck well, even on gravel or loose dirt. Altho tire squeal was fairly low, stress on outside tire in corners was vividly evident to a bystander. The 6.70 x 15 tires rolled under the rim alarmingly compared to the same size tires on a stiffer-sprung car.

The car with power steering was considerably more nimble, could be whipped thru tight turns without illusion of front-heaviness. Power steering didn't sacrifice directional control or create over- or understeer as is often the case with power steering. It was much easier to correct slides, drifts or misjudgments in the Nash with power steering.

PASSENGERS' COMMENTS

Exit and entry: Vertical cornerposts of wraparound windshield don't extend into passenger's angle of entry; doors open wide, have good stops, disclose a generous entranceway. Both cars were standard 4-door sedans, made no space concessions to hardtop styling.

Ride: It's hard to describe riding qualities of both cars without saying, "You've got to try them to appreciate the ride." Equipped with unbeatable reclining seats, Nash and Hudson offer passengers a ride that's free of bothersome shakes and rattles, but not opposed to body roll. Follow a 4-coil Hudson or Nash along a road that's sprinkled liberally with tarstrips, bumps and washboard stretches and watch the rear wheels, axle and differential case rise and fall; and keep an eye on the rear bumper of the car—it stays amazingly steady.

With the front seat tilted back to the position of your choice (ranging from easy chair slope to completely flat on your back) you find *any* trip untiring. And there's the advantage of the driver and/or passenger being able to stretch out and "take 10" or even sleep all night.

Rear seat passengers find the going smooth; with the front seat tilted backwards any more than 45 degrees, someone in back has to move. Tho legroom isn't as great as in many cars, you won't notice much difference; rear seat headroom isn't challenged until you get into a limousine.

CONSTRUCTION AND MAINTENANCE

How well they're put together: Both cars showed slight faults—a mismatched line between 2-tone painting on one car, misaligned ashtrays on both. (We found this on last year's cars also, then attributed it to early production.) The more important points—windshield molding, door panel construction, window mechanism, upholstery, fit of hood and of the rear deck showed no flaws

Servicing: You shouldn't run into complications here. Any competent mechanic can handle the A-M V8 with the proper tools and service instructions. Gas station attendants may be inclined to overlook the battery when you stop for gas and oil—the battery is up near the cowl, in a bad spot for a look under the cell caps. Sparkplugs are above the exhaust manifolds, but sheet metal of the fender wells crowds in toward the engine to complicate hasty removal of the center and rear plugs. Unit construction eliminates much periodic checking, for there are no body-frame bolts to loosen and cause latter-year failures.

OPTIONS

Other models: Hornet Special and Ambassador Special are available in a 2-door hardtop. Hudson comes in 1 sedan, Nash in 2. Standard shift, overdrive and Hydra-Matic are available on any model.

Equipment and accessories: Cars were alike in most respects. Hornet was less fancy, less costly Super model. It had no power steering or power brakes, but these items (or any accessories on the list) are available on the lower-priced cars. High on this list in this vacation month are the air mattresses and insect screens for camping. Moderately priced air conditioning is available in all A-M products, comes in a package deal with an excellent heater, doesn't take up any trunk space.

PERFORMANCE

(190-bhp engine)

ACCELERATION From Standing Start
0-30 mph 4.6 0-45 mph 8.5
0-60 mph 14.6
Quarter-mile 19.7 and 69.7 mph

Passing Speeds
30-50 mph 5.6 40-60 mph 6.9
50-80 mph 15.0

FUEL CONSUMPTION Used Mobilgas Special
Stop-and-Go Driving
16.2 mpg over cross-country course
12.0 mpg over traffic course
14.7 mpg tank average for 964 miles

Steady Speeds
21.1 mpg @ 30 19.0 mpg @ 45
16.1 mpg @ 60 13.7 mpg @ 75

STOPPING DISTANCE 151 feet from 60 mph

BRAKE FADE Slight on 4th stop from 60 mph
Complete fade on 6th stop with hard pedal action, no swerve
Recovered for effective stop on 7th stop; no further fade thru remainder of 12-stop test

TOP SPEED Fastest run 102.1 Slowest 98.8
Average of 4 runs 101.5

SPEEDOMETER ERROR Read 31 at true 30, 46 at 45, 61 at 60, and 77 at 75

SPECIFICATIONS

ENGINE: Ohv V8. Bore 3.50 in. Stroke 3.25 in. Stroke/bore ratio 0.928:1. Compression ratio 8.0:1. Displacement 250 cu. in. Advertised bhp 190 @ 4900 rpm. Bhp per cu. in. 0.76. Piston travel @ max. bhp 2654 ft. per min. Max. bmep 144.8 psi. Max. torque 240 lbs.-ft. @ 2000-3000 rpm.

TRANSMISSION: Standard transmission is 3-speed synchromesh with helical gears. Overdrive transmission is standard shift with planetary gearset. Automatic transmission is Hydra-Matic, 4-speed planetary gearbox with 2 fluid couplings.

RATIOS: Standard transmission: 1st 2.57:1, 2nd 1.55:1, 3rd 1.00:1, reverse 3.49:1. Hydra-Matic: 1st 3.97:1, 2nd 2.55:1, 3rd 1.55:1, 4th 1.00:1, reverse 4.31:1. Overdrive, 0.70:1.

REAR-AXLE RATIOS: Standard transmission 4.09:1, overdrive 4.55:1, Hydra-Matic 3.54:1 (3.07:1 optional).

STEERING: Number of turns lock to lock: mechanical 4.3, power 3.8 (Hudson); mechanical 4.0, power 3.6 (Nash). Overall ratio: mechanical 26.1, power 23.7. Type: worm and roller (mechanical and power).

WEIGHT: Test car weight (with gas, oil and water) 3846 lbs. (Hudson), 3894 lbs. (Nash). Test car weight/bhp ratio 20.2 (Hudson), 20.5 (Nash).

TIRES: 6.70 x 15 tubeless.

PRICES: (Including suggested retail price at main factory, federal tax, and delivery and handling charges, but not freight.) HUDSON HORNET SPECIAL Super 4-door sedan $2626, 2-door hardtop $2741. NASH AMBASSADOR SPECIAL Super 4-door sedan $2591, Custom 4-door sedan $2816, 2-door hardtop $2681.

ACCESSORIES: Flashaway Hydra-Matic $205, overdrive $109, power brakes $39 (available with automatic transmission only), power windows $110, power steering $108, radio $93, heater $80, air conditioning and heater $395, reclining seats and twin beds $23 (standard on custom models). Continental tire $125.

DIMENSIONS

- A FRONT OVERHANG **34.9**
- B WHEELBASE **114.3**
- C REAR OVERHANG **53.1** (**63.1** w/continental tire)
- D OVERALL HEIGHT **61.8** (**63.3** unloaded)
- E MINIMUM GROUND CLEARANCE **6.3** at muffler)
- F FRONT LEGROOM **42.5**
- G REAR LEGROOM **40.1**
- H FRONT HEADROOM **37.5**
- I REAR HEADROOM **36**
- J OVERALL LENGTH **202.3** (**212.3** w/continental tire)
- K OVERALL WIDTH **78**
- L FRONT SHOULDER ROOM **61.5**
- M REAR SHOULDER ROOM **61.3**
- N TRUNK CAPACITY **N/A**

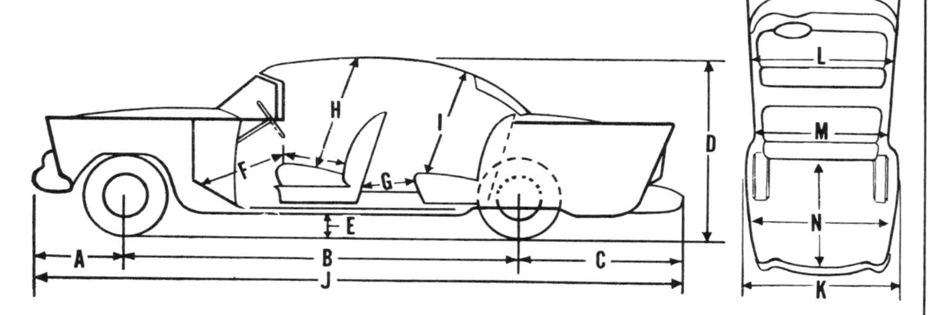

HUDSON ROAD TEST

(Continued from page 39)

5. The small bore gives the L-head combustion chamber an extended lease on life because compression ratios of 8 to 1 or slightly more can be used without sacrifice in volumetric efficiency. (About 7 to 1 is the absolute limit with a square engine—above that the transfer area proves restrictive).
6. The piston speed factor can be alleviated by good design and becomes negligible when used with an overdrive or a dual range hydramatic.

During the tests we found that valve-bounce occurred at an indicated 70 mph in 2nd gear, or 62.5 actual mph. This is 5500 rpm, equivalent to a piston speed of 4350 ft. per minute! During the performance tests we approached these speeds at least *75 times* and the little stroker was still running as smoothly and as quietly at the end as at the beginning.

GENERAL COMMENTS . . .

For a car featuring "step down" design, we thought 7 inches more overall height than a Studebaker hardly justifies calling this a low car. On the other hand, the seats are very comfortable and visibility is excellent. The driving position is good and the column shift control is positive, a feature that contributed to the excellent performance thru the gears.

The general quality of details and finish, both inside and out was well above average for this price category. The body was well sealed and wind noise was low.

Our advice to genial Johnny Lail, the Glendale Hudson dealer who made this test possible is to demonstrate this car and it will sell itself. To prospective purchasers we can only say: drive all 6 cars and may the best car win! . . . . JB •

AMERICAN MOTORS has chosen late Spring to bring forth the Hudson Hornet Special and the Nash Ambassador Special "$56\frac{1}{2}$" models. Both these cars are powered with American Motors' new 250-cubic inch, 190-bhp V8 engine, the first V8 ever built by the new firm.

The new engine has been placed in the short ($114\frac{1}{2}$") wheelbase chassis-body unit which is also shared by the Nash Statesman and the Hudson Wasp. What this means is that either a four-door sedan or two-door hardtop of $114\frac{1}{2}$" wheelbase is available with Nash body styling and trim. Engine choice is either the 130-bhp overhead valve, six-cylinder engine (used with single-barrel carburetor in AM's Rambler), or the new 190-bhp V8.

The same two models in Hudson trim are sold as the Wasp with a long-stroke, 120-bhp six, or with the new V8 as the Hornet Special.

The new short wheelbase V8s add what amounts to a brand-new car, (in fact *two* brand-new cars in name and trim: Hudson Hornet Special and Nash Ambassador Special) to the low medium-price field.

These cars are factory listed at **around $2700 and are competitive** with some of the "DeLuxe" models in the so-called low-priced field. In addition, they offer some unique features that add up to a transportation package seldom found in the low-priced car market.

First of all there is the large passenger compartment of the American-Motors welded unit body-frame, which offers more room than any cars short of the Buick Super, DeSoto, Cadillac and Packard. Furthermore, this space does not carry a penalty of great bulk and weight.

The Hudson Special weighs only 100 pounds more than the average V8 in the lowest price field.

The welded unit construction of body and frame, a feature unique to American Motors' products, makes the Hornet Special the quietest and most solid car in its price class. Most road vibration stops at the springs and cannot shake the rigid body-frame unit.

The short hood, which is sloped forward between the fenders, allows excellent forward vision. Another bonus is in easy parking and maneuvering.

The 190-bhp V8 is exceptionally well designed, developing more power than any other low-priced V8 in standard trim. This new engine is the smallest of U. S. V8s and uses a two-barrel carburetor, 8.0:1 compression ratio and takes regular gasoline. What this means is performance equal to any of the low-priced cars.

The Hornet V8 teamed with the newly refined version of Dual-Range Hydra-Matic (the same Transmission used on Cadillac, Olds and Pontiac) delivers performance as powerful as any car in the low and medium price field.

Summing up: An exceptionally comfortable, smoothly powered and well-engineered car offering more passenger space than any car within $1000 of its price. Only the unfavorable rate of depreciation and personal considerations can be held against the car. ●

HUDSON HORNET SPECIAL

Factory List Price: **$2741**

detroit's $1956\frac{1}{2}$ cars

☑☑☑☑☑ MEANS TOP RATING	HUDSON HORNET SPECIAL
PERFORMANCE ☑☑☑☑☐	Excellent; equal to all but the "special" powerpack engines in the low and medium-priced field. Smoother and quieter at all speeds than any other car in low-medium price field.
STYLING ☑☑☑☑☐	Basically a straightforward and pleasing design which is somewhat marred by over-enthusiastic application of trim details.
RIDING COMFORT ☑☑☑☑☐	Ride is firm, with some jiggling and bouncing at slow speeds on rough pavement. There is no sway and very little pitching.
INTERIOR DESIGN ☑☑☑☑☑	An outstandingly well-designed interior with plenty of elbow room and good vision all-around. There is more room than in any car of this price range, plus reclining seats and bed feature.
ROADABILITY ☑☑☑☑☐	One of the best in its class, although not as good as the pre-merger Hudsons. Does not roll, pitch or wallow in rough going. Some tendency to drift on sharp curves when lightly loaded.
EASE OF CONTROL ☑☑☑☑☐	Rates very well with power steering. Power steering is helpful but not essential, as this car is a good deal lighter than big Hudson V8. Power brakes not needed.
ECONOMY ☑☑☑☑☑	Should be about tops considering its size and performance.
SERVICEABILITY ☑☑☑☐☐	Fair to good. Engine compartment it crowded but engine is not too complicated. Spark plugs are accessible.
DURABILITY ☑☑☑☑☑	Seasoned, solidly-built American Motors body-frame construction plus a well engineered, low-stress engine and proved Hydra-Matic should add up to a very long-wearing and trouble-free package.
VALUE PER DOLLAR ☑☑☑☑☐	Much better than the rest of the line (Hudson or Nash) because of combination of economy, comfort and performance.

'57 HUDSON and NASH

A flatter roof and 14-inch wheels lower the '57 Hudson (right) two inches below the '56.

by Don MacDonald

THE AMERICAN MOTORS TWINS, Hudson (above) and Nash, get more alike than ever before as their design matures with the passing years. Past variety in model names, wheelbase and engine options has been eliminated. Each offers a 121¼-inch-wheelbase Super (less deluxe) and Custom (more deluxe) models powered by a home-grown V8 of 255 horsepower. The 190-horsepower version of this engine was well-proved by service experience with the "Special" series introduced in mid-1956, and still serves as a potent option on the 1957 Rambler (see page 43).

The 65 numerical jump in horsepower is gained from the combination of a half-inch larger bore, one unit greater compression ratio (now 9.0 to 1), two more barrels in the carburetor, plus breathing refinements. Standard dual exhausts help at the rear wheels but are not normally counted when rating engine horsepower.

Both cars are two inches lower than last year's equivalent models, making them an even five feet at centerpoint. Reduction in height was achieved by flattening the roof and changing to 14x8.00 tires.

Nash shows signs of receiving the major share of the family's facelifting budget; it takes a sharp eye to detect any difference between '56 and '57 Hudsons, or Ramblers for that matter. Nash fender fronts, formerly occupied by vast parking lights, now sport a genuine dual headlight system, the first in the industry.

Everyone else is marking time, waiting on nine recalcitrant states to legalize the vital improvement. Lincoln, for example, is temporarily installing "fog" lights where the second part of the dual system should be. Mercury and Chrysler products will offer the expensive (to them) optional grille. Hats off to Nash for plunging ahead; we doubt if any owners will be penalized for their courage.

The single filament lower pair of lights combined with the "city driving" filaments in the upper pair flood a clear highway at 150 watts output. The remaining filaments in the upper pair are used for passing.

Optional new type Hydra-Matic and overdrive are continued on both cars, with three-speed standard. Ultramatic has disappeared with the abandonment of the Packard-supplied engine. The peculiar (to Nash and Hudson) pull-on-the-gear-shift-lever starting has been much improved by incorporating a vacuum cutout designed to prevent inadvertently grinding the starter motor when the engine is running. All six-cylinder models have been discontinued. Body choices, in a variety of new colors, are limited to four-door sedans and two-door hardtops.

Inboard lights on '56 Nash (left) have given way to the first duals. They're standard, come what may.

Price range (Factory list price)
$2,520 (Hornet Super four-door sedan)
to $2,781 (Hornet Custom two-door hardtop)

HUDSON

By JAMES WHIPPLE

IN TUNE WITH American Motors' sales policy for 1957, Hudson Motors Division has eliminated the somewhat confusing array of models and lowered prices.

In recent years there have been too many Hudsons (and Nashes) competing against each other for their potential market.

In 1956 for example you could get the same basic Hudson four-door sedan body shell on 114-inch wheelbase with either an L-head six cylinder engine, or American Motors' 190 bhp V-8, or on a 121-inch wheelbase with famous, husky "Hornet" L-head six or with the Packard-built V-8 rated at 220 bhp. To add to the buyer's bewilderment all these models were available in standard ("Super") and deluxe ("Custom") trim.

This year Hudson has cut out all models but one, the Hornet, which has a 121-inch wheelbase and is powered by a 255-bhp version of American Motor's newly-designed V-8 engine. Like last year there are two body types; a four door sedan and a two door hardtop. These body types come in Super and Custom trim as before.

Transmission options include the "new" Flashaway Hydra-Matic with its smoother fluid clutch automatic shifting or standard synchromesh and standard synchromesh plus overdrive.

At $2750 factory list plus Federal tax and handling charges for the Super four door sedan, Hudson is in direct price competition with Pontiac Super Chief, Mercury Monterey, Dodge Royal V-8, Buick Special, Olds 88 and De Soto Firesweep.

In wheelbase, weight and horsepower, Hudson comes close to the other cars in this price group. At 3,660 lbs., the Hornet weighs within 50 to 100 lbs. of the others in the group except for the 4,000-lb. Buick and Olds. As far as power is concerned, Hudson's 255 bhp is about average.

Although the car we tested was equipped with Hydra-Matic transmission and a low ratio ("high-geared") 3.15 to 1 rear axle, performance was quite brisk with an informal time of 10.7 seconds from standing start to an indicated 60 mph.

With the three-speed stick shift transmission which is standard equipment, Hudson furnishes a 4.1 to 1 axle ratio which should make for a very rapid acceleration in the low and medium speed ranges. For anyone who's remotely interested in gasoline economy at cruising speeds, overdrive would be recommended as a valuable option with the three speed transmission.

Driving the '57 Hudson V-8 is a good deal more enjoyable than last year's Hornet V-8. Principally because handling has improved considerably. Two factors are responsible

HUDSON is the car for you

if... You appreciate one of the roomiest, most comfortable and practical interiors of any car on the market.

if... You like a car that's absolutely rattle-free at all times and rock solid at high speeds on rough pavement.

if... You want powerful performance combined with the economy potential of Hydra-Matic or overdrive transmissions.

if... You are willing to accept higher than average depreciation and consider long term ownership as the price of Hudson's unique advantages.

HUDSON SPECIFICATIONS

ENGINES	
Bore and stroke	4.00 in. x 3.25 in.
Displacement	327 cu. in.
Compression ratio	9.1:1
Max. brake horsepower	255 @ 4700 rpm
Max. torque	345 @ 2600 rpm
DIMENSIONS	
Wheelbase	121.3 in.
Overall length	209.3 in.
Overall width	78 in.
Overall height	60.4 in.
TRANSMISSIONS	Standard synchromesh, overdrive, Hydra-Matic

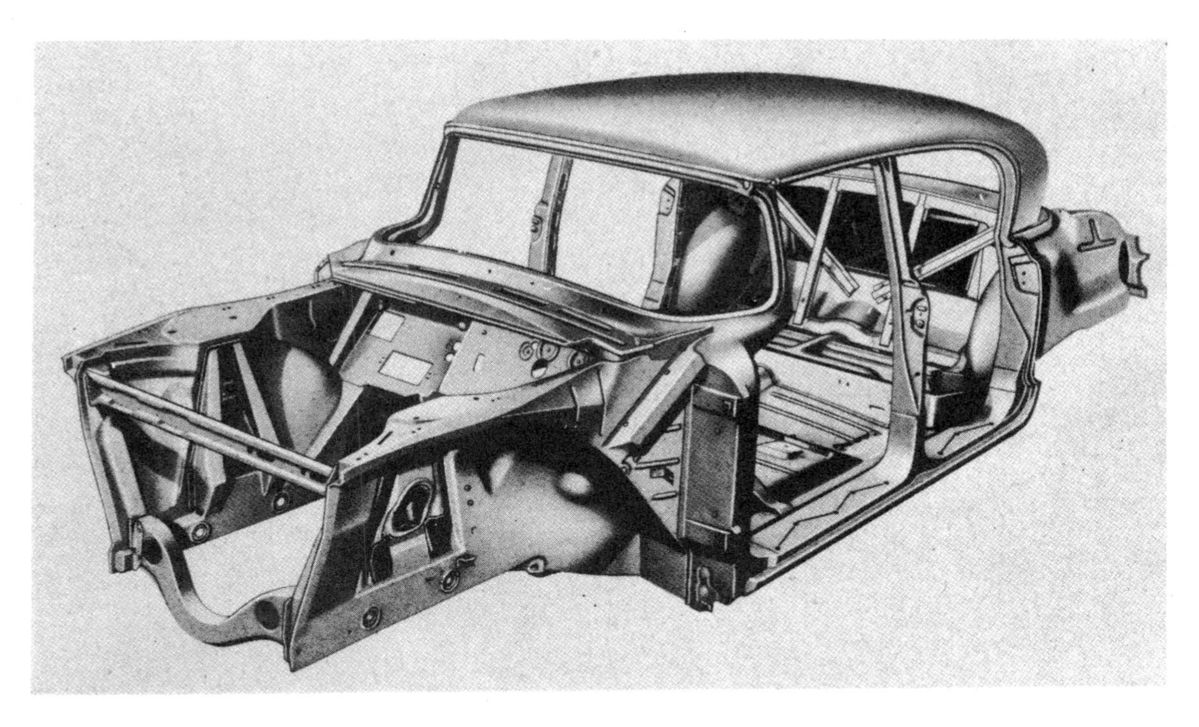

American Motors' single unit body construction, used by Hudson, incorporates 9,000 separate welds. Body and frame form one rigid, integrated unit. This type of construction is rattle-free, exceptionally durable.

for this. First, the use of American Motors' more efficient and lighter V-8 engine has reduced the sedan's weight by about 170 lbs. Almost all of this avoirdupois represented cast iron of the beefy Packard-built V-8 which displaced 352 cubic inches, yet developed only 220 bhp at 4600 rpm. This compared to the 57's new "over-square" (bore 4" stroke 3¼") AM V-8 engine of 327 cubic inches, which develops 255 bhp at 4700 rpm and at a lower (9.0:1) compression ratio.

The second factor is the replacement of the front wheel kingpin and bushings setup with a one-piece forged-spindle carried on anti-friction bearings. This has reduced turning effort and, coupled with the reduction in front-end weight, has permitted a lower steering ratio (18 to 1 instead of 22 to 1).

The result is easier steering and less wheel cranking. The reduction in front-end weight has given more equal weight distribution between front and rear. The Hornet V-8's balance is now approximately the same as last year's Hudson Hornet Six which was a better-than-average car as far as handling and roadability were concerned.

Most drivers are still going to want their Hornets with power steering, which was an absolute necessity on the nose-heavy 1956 V-8. Although the '57 Hornet V-8 is a much easier steering car, the weight is still considerable and the new broader treaded 8.00 x 14 tires make wheel cranking a chore when parking.

The new steering setup and balance make driving the 'Hudson an enjoyable occupation. As always the unit body and frame, built up of pressed sheet steel reinforced with box sections and held together by 9,000 electric welds, remains solid as a bank vault over any road surface. Sound-deadening materials and vibration-blotting rubber insulation at the suspension block out road noise and vibration, while the powerful V-8 is less audible than the engines of most cars.

This one-piece body frame structure, (American Motors calls it "Unitized,") gives the Hudson a solid quiet ride at all times which makes the car feel as if it were hundreds of pounds heavier than its 3,600 lbs. The unitized construction is particularly welcome on the hardtop two-door "Hollywood" model which is the most squeak- and rattle-free car of its kind.

In addition to the safety advantage of unitized construction (it resists the crushing force of impact better than separate body and frame design), incorporating the function of the chassis frame into the body itself eliminates extra weight. This in turn permits a larger car for a given weight. Hudson's seats are 64½ and 65 inches wide at hip and shoulder dimensions and legroom is exceptional for a car of 121-inch wheelbase.

The roofline has been lowered 1½ inches as American Motors follows the industry's trend toward lower and lower cars. The lowering process was achieved through redesigning the roof panel and making it much flatter. Interior headroom has been reduced 1⅛ inches in the front seat and only ¾ of an inch in the rear which leaves headrooms of 36⅜" and 35¼" respectively. These dimensions are still above average for the industry and will pose no problem for Hudson owners who like to ride with their hats on.

Hudson riding qualities have been improved over last year's car principally by the switch to 14-inch wheels with larger tires and lower air pressures. The lower pressure tires permit better absorption of shock.

With the improved balance and

Behind massive Hudson grille is new V-8 engine. This power plant develops 255 horsepower. An AM product, it comes with a four-barrel carburetor and has dual exhausts as standard equipment.

weight distribution, the Hudson is even less prone to bobbing and pitching on uneven roads, and shock absorber control or rebound is excellent.

Moderate-sized bumps still create some well-muffled shock, however. This seems to be one of the problems inherent in an all coil-spring suspension. In this type of suspension, the coils must be a compromise between softness to absorb bump shock and stiffness to prevent rolling, pitching and swaying.

The advantage of a coil-spring rear suspension found on Hudson (Nash, Rambler and Buick) is that the rear axle is rigidly located by torque tube and track bar so that there is no wheel hop in acceleration or uncontrolled skittering when cornering hard on rough pavement.

This rigid alignment of rear wheels (as well as front) plus the absolutely inflexible unit body-frame makes the Hudson one of the most precise handling, and consequently pleasant and relaxing to drive, cars on the road.

SUMMING UP: Hudson is an exceptionally roomy, solidly built and comfortable car with performance that's smooth and powerful in all speed ranges. ●

HUDSON CHECK LIST

5 CHECKS MEAN TOP RATING IN ITS PRICE CLASS

PERFORMANCE	Powerful, smooth performance with Hydra-Matic is a bit above average for Hudson's price class. New V-8 operates quietly except under extreme acceleration.	✔ ✔ ✔ ✔
STYLING	Hudson is a classic case of the problems encountered in attempting to make a basically clean-lined and well proportioned car look "different" without major changes. Body lines are somewhat cluttered, while front and rear areas are overloaded with trim.	✔ ✔ ✔
RIDING COMFORT	Overall riding qualities are excellent. Softer tires blot up surface irregularities and stable, well-balanced chassis prevents pitching and swaying. Somewhat stiff coil springs transmit some of the impact of medium-sized bumps such as railroad crossings.	✔ ✔ ✔ ✔
INTERIOR DESIGN	Hudson's body is exceptionally roomy in all important dimensions. Seats are at a comfortable height and well upholstered. Seat reclining (and convertible car bed) feature is a low cost and well worthwhile extra for touring. Instrument panel is well located and easy to read.	✔ ✔ ✔ ✔
ROADABILITY	Redistribution of weight, and new, lower steering ratio and front wheel mounting make Hudson a good road car that stays in the groove and steady on curves and rough going. Car is very stable at high speed. Overall handling is excellent—much improved over '56 Hudson V-8.	✔ ✔ ✔ ✔
EASE OF CONTROL	New bearing-mounted front wheel spindles and lower steering effort (and gear ratio) make Hudson easy to control on the open road. Power steering is recommended for buyers who do a lot of city driving.	✔ ✔ ✔ ✔
ECONOMY	Hudson offers an unusual economy potential for a car of its power and weight with choice of economical Hydra-Matic or overdrive transmissions.	✔ ✔ ✔
SERVICEABILITY	Hudson's big, wide V-8 is crammed deep into the narrow engine compartment of the American Motors one-piece body and frame unit making servicing of such units as power steering difficult at best.	✔ ✔ ✔
WORKMANSHIP	Overall average of quality is high. Paint trim and upholstery all up to snuff for Hudson's price class.	✔ ✔ ✔ ✔
VALUE PER DOLLAR	Good for the owner who is prepared to spread the car's high initial depreciation over several years (or many miles) of ownership. Hudson is an excellent value under these conditions.	✔ ✔ ✔

HUDSON OVERALL RATING . . . 3.6 CHECKS

LEVER on steering column allows driver to select desired gear range; his choice is indicated by letters on crescent-shaped dial over column.

Briefly, this is how the previous model and the current car compare:

	PREVIOUS	CURRENT
0-50 m.p.h.	10.8 sec.	9.7 sec.
0-90 m.p.h.	39.4 sec.	42.1 sec.
Max. speed	102.0 m.p.h.	100.3 m.p.h.
Consumption	18.0 m.p.g.	16.6 m.p.g.

Now let's look at the transmission. Fundamentally it is a four-speed and reverse mechanism operated automatically by the combination of throttle setting and engine revs.

Just like all the others? — Not quite . . .

YOU Control it

The designers of this transmission have actually admitted that its mechanical brain is not superior to the human equipment.

They have built-in a degree of overriding driver control that puts the Hydramatic unit in a class above any other automatic unit I have tested.

A selector lever on the steering column moves through four positions: N (Neutral), D4 (Drive range with high 4th gear for fast cruising in operation), D3 (Drive range with only the three lower ratios working), L (Low range), and R (Reverse).

So, by being able to select between D3 and D4, the driver can decide which range of gears is best suited to his driving conditions.

HUDSON HYDRAMATIC

Driver keeps charge of this car instead of playing stooge to the automatic drive; a good thing, says Bryan Hanrahan

THIS road test is of the first automatic Hudson Hornet to be regularly imported to Australia.

By this I mean that it's no longer the sort of car you have to wangle into the country with the help of a rich uncle in America, or your own well-lined overseas dollar account.

Apart from the very interesting Hydramatic transmission and some styling changes it is mechanically similar to the previous model, which I tested for **Modern Motor** exactly twelve months ago.

Performance is much the same, too, except that the acceleration curve shows the usual effects of the incorporation of automatic transmission.

Extra friction and slip in the Hydramatic unit give the car slightly better acceleration up to 70 m.p.h., but it is slower from there on, up to a slightly reduced maximum of 100.3 m.p.h.

Fuel consumption is slightly greater, for the same reason.

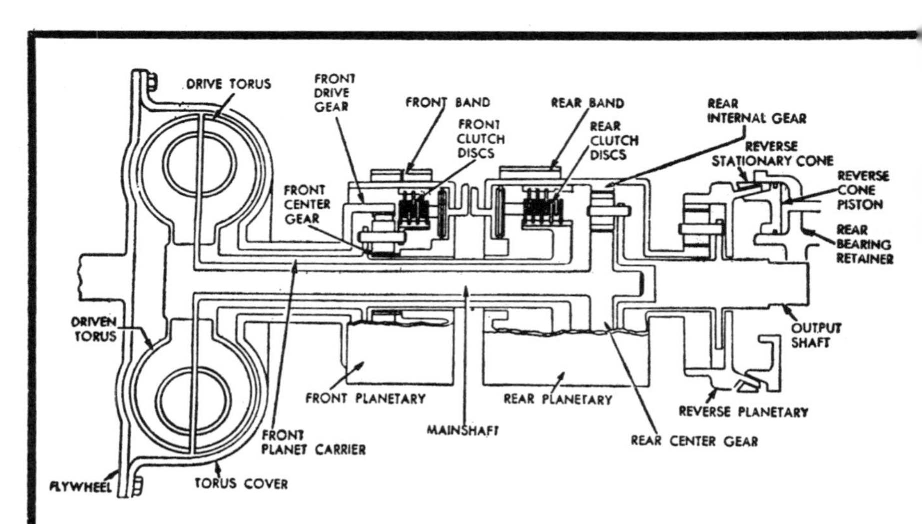

DIAGRAM of Hudson's Hydramatic system, showing how drive is coupled by planetary gears and clutches; a cone clutch gives reverse.

D3 gives better acceleration at low speeds and is designed to cope with heavy traffic. D4 brings in an overdrive top gear for fast cruising.

In D4 the transmission automatically shifts down to third when road speed falls below 60 m.p.h. D3 goes into second at less than 20 m.p.h. You can exceed these figures in the lower gears when the accelerator demands it—but once the car speed falls below the limiting figures, down a ratio you go.

(These limits are additional to the normal upper and lower limits of each gear's speed range. Most transmissions, when lightly loaded, will stay in top down to a crawling pace: not until the throttle setting is varied does a change-down occur. The upper limits are fixed but are only reached with extreme low revs and wide-open throttle; an upward change is automatic at practically any speed as soon as throttle opening is decreased.)

Kick-down Switch, Too

The other overriding device is a kick-down switch on the accelerator. If you press the pedal flat to the boards, the transmission will go straight into the lowest gear allowed by the speed at which the car is travelling.

Example: you are tittling along in top and urgently need some herbs.

Most automatics will go into the next lowest gear—after a slight delay —as the throttle is opened wide. The box has to "think" for itself.

MAIN SPECIFICATIONS

ENGINE: 6-cylinder, s.v.; bore 3 13-16in., stroke 4½in., capacity 308 cu. in.; compression ratio 7.5 to 1; R.A.C. rating 34.88 h.p.; maximum b.h.p. 160 at 3800 r.p.m.; downdraught double-choke carburettor; mechanical fuel pump; 12-volt ignition.

TRANSMISSION: Hydramatic four-speed unit; hypoid bevel final drive.

SUSPENSION: Front independent by long coil springs and wishbones; rear, semi-elliptics; telescopic hydraulic shock-absorbers all round.

STEERING: Worm-and-roller, turning circle 39ft. 7in.; 4 2-3 turns lock-to-lock.

WHEELS: 15in., with 7.10 by 15in tyres; pressures, 24lb. all round.

BRAKES: Hydraulic, servo-assisted.

CONSTRUCTION: Unitary.

DIMENSIONS: Wheelbase 10ft. 1¼in., front track 4ft. 11in., rear 5ft.; length 17ft. 5¼in., width 6ft. 6in., height 5ft. 2½in.; clearance 8¼in.

DRY WEIGHT: 3677lb. (just under 33cwt.).

FUEL TANK: 20 gallons.

PERFORMANCE ON TEST

CONDITIONS: Hot and dry, no wind; smooth bitumen, two occupants, premium fuel.
MAXIMUM SPEED: 100.3 m.p.h.
STANDING quarter-mile: 19.6s.
FLYING quarter-mile: 98 m.p.h.
ACCELERATION through gears of automatic transmission: 0-30, 4.5s.; 0-40, 7.1s.; 0-50, 9.7s.; 0-60, 15.2s.; 0-70, 22.0s.; 0-80, 31.1s.; 0-90, 42.1s.

BRAKING: 35ft. 2in. to stop from 30 m.p.h.

FUEL CONSUMPTION: 20 m.p.g. at steady 30 m.p.h.; 18 at 60, 16.6 overall, including all speed and acceleration tests.

SPEEDOMETER: Accurate at 30 m.p.h., 4 percent fast at 60, 7 percent fast at 90.

PRICE: £2980 including tax

FLASHBACK—In case you haven't the issue for March 1956, here are the relevant test figures for the previous normal-drive Hudson, so you can compare them. MAXIMUM SPEED: 102 m.p.h. STANDING quarter-mile: 19.8s. FLYING quarter-mile: 97.8 m.p.h. ACCELERATION from rest through gears: 0-30 4.9s.; 0-40, 8.0s.; 0-50, 10.8s.; 0-60, 15.4s.; 0-70, 21.0s.; 0-80, 29.6s.; 0-90, 39.4s. FUEL CONSUMPTION: 22 m.p.g. at steady 30 in top gear; 21 m.p.g. at steady 60 in top overdrive; 18 m.p.g. overall, including all tests

With the Hydramatic unit, you operate the kick-down switch and INSTRUCT it to select the lowest permissible ratio. In the D4 range this will often take you down two, sometimes even three, ratios from top.

The "brain" is overridden and it can't waste time "thinking."

Result: YOU select the right gear for maximum getaway—and, in the Hudson, get away you certainly do.

This return to human control is most welcome. Anything that intrudes mechanical "thought" into the realm where a driver's intelligence and skill are required is an extremely bad thing.

Fully automatic transmissions think in terms of engine revolutions and throttle opening ONLY—they know nothing about the pedestrian you may need full-torque to avoid, low-gear braking you may need if your wheel brakes fail, or if you're driving on a slippery surface.

But don't let me put lazy drivers who adore automatic transmissions off the Hudson unit—you don't have to use the overriding devices unless you want to.

The other advantages of the Hydramatic transmission are that the car can be push-started if motor or battery are playing up; and in case of breakdown (other than serious transmission trouble) it can be towed with all four wheels on the road.

To appreciate these points fully, I recommend you read a book put out by Shell on the servicing of automatic transmissions. Many can neither be push-started nor towed with the rear wheels on the deck.

About the Car

Well, with the reservations made in the last Hudson road test, I thought the current model a superior American-style car.

ENGINE is the well-tried six-cylinder side-valve, powerful yet durable.

I won't go through all the details again but mention only new points, good and bad.

The gear-selector lever is insufficiently protected from accidental engagement of L or R. You are supposed to lift it before moving it to select these ratios, but the precaution isn't very effective.

The main controls, particularly the steering, are pleasantly light; but the accelerator is still set too high on the toeboard for comfort.

Some of the minor controls, such as the wiper knob and the windscreen-washer button, are positioned for left-hand drive. Since the seats are over five feet wide, this is extremely awkward—not to say disgraceful in a car costing nearly £3000.

Windscreen reflections are very bad.

Handling Characteristics

Despite its huge bulk (17ft. 5$\frac{1}{4}$in. by 6ft. 6in.) the Hudson is easy to handle under normal circumstances.

A reduced turning circle (39ft. 7in. instead of 42ft. 8in.), coupled with the light but low-geared steering, helps here.

Fast or slow, on the straight the Hudson rides better than any other American car I know. There's no floating—even a suspicion of firmness.

Cornering is a different matter.

What the precise fore-and-aft weight distribution is, I couldn't find out. But an awful lot of weight seems to be thrown on the front wheels.

The car simply wants to go straight ahead.

Sports-car performance is backed by magnificent servo-assisted brakes. They do tire with repeated application, but again, the Hudson beat most of its compatriots in this respect.

The handbrake is lousy and the transmission hasn't got a positive mechanical lock for parking. You are advised by the handbook to leave the car in L or R—which presumably means that the car can't exceed 5 m.p.h. if it runs away.

That's about the speed at which the fluid in the transmission would couple up to the dead motor.

A serious defect, indeed.

An extra good point to my mind is that the 160 b.h.p. Champion Six engine is a low-revving side-valve job —the type of engine which established the American reputation for longevity.

Ignition is now 12-volt—another welcome improvement.

Appearance is a matter of taste; finish is quite good. Generally, this new Hudson is one of the few automatics that appeals to me. • • •

HUGE boot matches the car's size; there's lots of room everywhere.

12,000 Miles Later

SIA's Hudson staff car almost halves Mercedes' per-mile cost despite shameful gas mileage.

SIA's BERZOOKY EDITORS intend to find out—can a 1952 Hudson Hornet on Medicare serve as this magazine's day-in, day-out workhorse?" We asked that question back in June 1972 [*see* Low-Bucks Staff Car, *SIA #11*]. Now, 18 months later, we have an answer.

The answer is yes, it can, but no, the car and the experiment didn't really come up to all expectations. I had hoped for more reliability and better economy. The Hornet's shop downtime and repair costs were higher than expected (a total of about six weeks down and $360.63 cash outlay), and fuel mileage—at 10.3 mpg—almost put me in the hospital.

Yet despite these disappointments, the dowager Hornet still lives and breathes fire, and it *did* beat the national average for cost-per-mile economy. All considered, it's still cheaper to drive a car like this than most newer ones. I imagine that if you were to pick a collectable car that gave better gas mileage and needed fewer repairs, you could get cost-per-mile down below a Volkswagen's.

The typical American driver, so the AAA tells us, spends $1647 a year to support his automotive habit. That's for fuel, depreciation, insurance, maintenance, license, taxes, and the rest. This averages out to 16.5¢ a mile based on 10,000 miles of annual driving. Keep the AAA figure in mind: 16.5¢ a mile. It's official ("") and a good theoretical yardstick, but remember, too, that in actual practice a car's cost per mile varies according to its size, popularity, your driving habits, and where you live.

I drive, as you know, the '52 Hollywood coupe you see pictured here, and I live in Stockton, Calif. Longtime SIA readers will recall that I bought this car in June 1971 from an 81-year-old Stocktonian. The odometer showed just under 76,000 miles at that time. I paid $350 for the car, and it was in good, good condition.

I've always had a soft spot for Hudsons, because my dad drove them from 1939 through the bitter end in 1954. Besides bringing back all those fond childhood memories, I consider Hudsons extremely well built and over-engineered. That opinion hasn't changed—I'm still prejudiced in favor of Hudsons.

Before putting the Hornet on the road back in 1971, I spent another $934.82 by way of refurbishments. I use the word refurbishments because this wasn't a restoration. The Hornet is very presentable, but it's never been in anything near show condition, nor was the idea to turn it into a prizewinner. My plan has always been to make this car a guinea pig—an experiment to see if a 20-year-old (now 22-) car can survive everyday driving with reliability and economy. It had to do that, I felt, without any coddling.

My initial reasoning went like this: Since a Hudson can't depreciate, and since it's likely to *appreciate* as time passes, it and many cars like it should logically make sound, economical transportation. The question, then, is, are such cars reliable, dependable, comfortable, and usable on an everyday basis? I'd like to explore my own log book and conclusions. You can then make up your mind.

A Basis For Comparison

ROAD & TRACK Magazine publishes a series of Extended Use Reports, and these have always interested me a great deal, especially since I participate in a similar series for POPULAR MECHANICS called Owners Reports. Both involve living with a car for a long period of time. Most magazine road tests, including SIA's driveReports, take only a day or so for the editors to complete, so they're all right for quick impressions, but they never reveal all of a car's quirks—not the things a real car owner uncovers after a year or two behind the wheel.

I felt it appropriate, then, to use one of R&T's Extended Use Reports as the model for my own experiences with the Hornet. I picked as my basis of comparison the car that R&T editor Ron Wakefield owned for 24,000 miles, a 1973 Mercedes-Benz 280 4-door sedan. Laugh not!

There's a difference, true, between driving up to someone's front door in a new Mercedes and doing the same in a 22-year-old Hudson. Even so, they're more alike than not. First and least important, both use in-line 6s and unit bodies. Second, they weigh nearly the same—the Hornet is only 250 pounds heavier despite being considerably bigger in every external dimension. Third, Mercedes and Hudson share similar longstanding reputations for quality, handling, performance, driving comfort, and as long-distance road cars. Fourth, in their day, both left indelible marks in racing (though not the Mercedes 280).

Their greatest difference comes in price, the upshot being that the Mercedes cost 22.2¢ a mile to drive and the Hudson cost 11.6¢ a mile. The Hornet, then, fell sweetly below the national AAA average, but in fairness, it didn't do so well as R&T's Opel Manta, which averaged 10.6¢ a mile over a 12,000-mile Extended Use Report. Let's continue with our Mercedes/Hudson comparison by category.

Horrible Gas Mileage

The Mercedes averaged 16.4 mpg for R&T's 24,000 miles of driving. That's quite a bit better than the Hornet's shocking 10.3 mpg. Yes, 10.3 mpg stinks, but the figure *is* accurate, unfortunately, and if it's any consolation, it came with regular-grade fuel, most of it around 30¢ a gallon at the time.

When mechanic Jim Holmes and I overhauled the Hornet's engine three winters ago, I swapped the stock manifold and 2-barrel carburetor for Hudson's Twin-H-Power setup. More glamor, I figured, and maybe even some better acceleration and gas mileage. The Twin-H manifold came from a local wrecking yard, but the Twin-H Carter W-1 *carbs* were new-old stock, still in their boxes. I bought them by mail from a fellow in New York, knowing full well that they were meant for the 1954 Twin-H, not the '52. In 1952, Twin-H upped Hornet horsepower from 145 to 160 bhp gross. By 1954, standard Hornets were pulling a base 160 bhp without Twin-H, and the dual carbs supposedly added another 10 bhp.

Now, whether having the wrong W-1 carbs reduces gas mileage or whether it would be that low anyway I don't know. I do know that MOTOR TREND, in their test of a 1953 Twin-H Hornet, averaged 13.3 mpg in city driving, and ROAD & TRACK averaged 9.5 mpg for their 1952 Hornet! That's already bad, especially for factory-tuned test cars, but then when I consider that I did drive on the freeway at 70-80 mph (this was before the present 55-mph limit, which I now observe religiously), and that I make lots of quick trips in town, I believe the 10.3-mpg figure stands to reason.

I realize that my Hudson-driving friends will gnash their teeth at the 10.3-mpg admission. Sure, a Twin-H Hornet can do better. Featherfeet might get 16-17 mpg. If you calculate mileage on only one tank of gas, you can get any figure you want. My calculations span 12,000 miles of every type of driving, though, some of it fast, some pokey, so this check is

Happy the editor who can find a Saturday to play. Among the Hornet's blessings is plenty of underhood workspace. Fender panel unbolts, lifts out in 10 minutes, gives fine access to tappets.

Hornet convertible parts car lives behind garage, recently donated its rear axle for a transplant.

Comparing Pocketbooks— SIA's Hornet vs. R&T's Mercedes	**1973 Mercedes-Benz 280, 2 years & 24,000 mi.**	**1952 Hudson Hornet, 1 year & 12,000 mi.**
Initial price	$9,377.00	$350.00
Refurbishing expenses	---	934.82
Repairs & replacements	268.00	360.63
Routine maintenance	230.00	39.98
Gasoline	600.00 (2 yrs)	771.10 (1 yr.)
License & taxes	306.00 (2 yrs)	32.00 (1 yr.)
Insurance (see below)	548.00 (2 yrs)	122.82 (1 yr.)
Total expenses	**$11,329.00**	**$2,611.35**
Resale value (est.)	6,000.00	1,250.00
Cost of driving	5,329.00 (2 yrs)	1,395.71 (1 yr.)
Overall av. cost per mile	**22.2¢**	**11.6¢**

Note: Mercedes-Benz figures courtesy **Road & Track** Magazine, Oct. 1973. Insurance on the Mercedes was 15/30/10 with $100 deductible collision & uninsured motorist. Insurance on the Hudson was (is) 25/50/10 with no collision and 15/30 uninsured motorist.

Hornet shines in roadability, handling, and as long-distance mover. As a road car it rivals Mercedes. Fatigued axle (right) snapped at 85,695 miles.

Hudson Hornet *continued*

definitely accurate. Whether it's typical I don't know. I suspect it is. Different jets or metering rods might help.

When I get a chance, too, I want to re-install the old 2-barrel stock manifold and run another 12,000-mile check. We'll see.

Routine Maintenance—How Much?

I change oil every 1000 miles and hand-lube the Hornet myself at 2000-mile intervals. Oil by the case costs 39¢ a quart (Standard 30-W detergent), as opposed to 80¢ at the station. Chassis grease costs 29¢ a tube, enough for a dozen lubes. I own a pair of J.C. Whitney steel lift ramps ($14 five years ago) and a cheap creeper ($9.95), which I figure have paid for themselves by now.

I installed an AC oil filter after we overhauled the engine. I change filter elements every 5000 miles. The Hornet burns no oil between changes, but it does drool a little from the rear main seal. R&T's Mercedes also had a leaky main seal, but this was fixed under warranty. The Hudson, being slightly off warranty, continues to leak, because dropping the pan makes replacement a complicated repair.

In summary, the Mercedes' routine maintenance was $115 a year. The Hudson's $39.98.

How About Repairs?

Repairs are another story. You'll remember that I'd nursed the old 4-speed Hydra-Matic along for several thousand miles in 1972 by changing fluid and adjusting the throttle rod for smoother shifts. I managed to sidestep an overhaul for a while, but eventually the 2-3 shift got rough again, and at 80,660 miles, this roughness forced me to the transmission shop. It was tearing the devil out of U-joints and the rear axle, causing all sorts of play in the drivetrain and, as I would find out 5000 miles later, this play had already weakened the right rear axle shaft so badly that it would eventually snap.

I took the car to the only good transmission shop here in town, and for $211.68 they put the Hydro back to rights. Or at least partially back, because I had to return twice more to let them correct a sticky shift from forward to reverse and also a hesitation going into fourth under load. In all, transmission repairs accounted for about 10 days of downtime, including one weekend.

308-cid L-head 6 with optional 2-carb setup and Hydra-Matic averaged only 10.3 mpg in 12,000 miles. Performance is adequate but not startling. Everything would improve with overdrive.

Major Repairs in 11,590 Miles

Date	Mileage	Repair	Cost
6/19/72	77,594	Clean & resolder radiator	$26.50
9/13/72	78,901	Muffler, replace	19.37
11/ 7/72	80,660	Overhaul Hydra-Matic	211.68
12/14/72	80,695	Center driveline bushing	6.00*
1/ 4/73	81,230	Balance one front tire	2.00
1/ 4/73	81,230	Set up brakes	3.00
2/ 1/73	82,612	2 steering bearings	6.87*
9/ 1/73	85,502	Gasket set & valves	14.75
9/ 4/73	85,502	Valve job, labor	10.00
9/12/73	85,695	2 bushings for axle	5.21
9/12/73	85,695	Machinework on bushings	12.50
9/12/73	85,695	Install new 3rd member, labor	42.75
		Total	**$360.63**

***Asterisk** denotes no labor or installation charge.

specifications

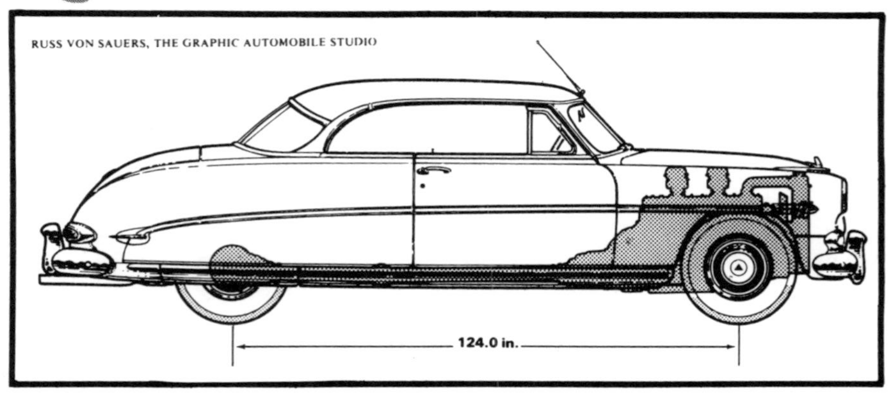

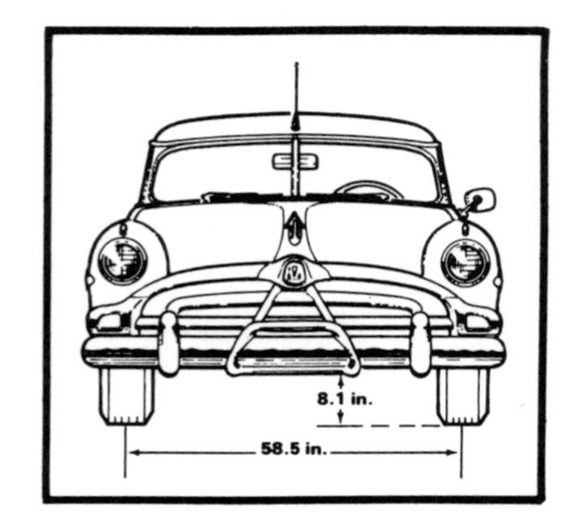

1952 Hudson Hornet 7B Hollywood hardtop coupe

Price when new	$2790 f.o.b. Detroit (1952).
Options	Radio, heater, Hydra-Matic, Twin-H-Power, day/night mirror.

ENGINE

Type	In-line L-head 6, water-cooled, cast-iron block, 4 mains, full pressure lubrication.
Bore & stroke	3.8125 x 4.5 in.
Displacement	308 cid.
Max. bhp @ rpm	160 @ 3800.
Max. torque @ rpm	260 @ 1800.
Compression ratio	7.2:1.
Induction system	Twin 1-bbl. carbs, mechanical fuel pump.
Exhaust system	Cast-iron manifold, single muffler and resonator.
Electrical system	6-volt battery/coil.

CLUTCH

Type	Non

TRANSMISSION

Type		Hydra-Matic 4-speed automatic, torque converter, planetary gears.
Ratios:	1st	3.82:1.
	2nd	2.63:1.
	3rd	1.45:1.
	4th	1.00:1.
	Reverse	4.30:1.

DIFFERENTIAL

Type	Hypoid, Hotchkiss drive.
Ratio	3.07:1.
Drive axles	Semi-floating.

STEERING

Type	Worm & roller.
Turns lock to lock	5.75.
Ratio	20.4:1.
Turn circle	42 ft.

BRAKES

Type	4-wheel hydraulic drums, internal expanding, with mechanical reserve.
Drum diameter	11.0 in.
Total lining area	158.7 sq. in.

CHASSIS & BODY

Frame	Unitized with body, perimeter type with bolt-on front subframe.
Body construction	All steel, welded.
Body style	2-dr., 6-pass. hardtop coupe.

SUSPENSION

Front	Independent unequal A-arms, coil springs, tubular hydraulic shock absorbers.
Rear	Longitudinal leaf springs, tubular hydraulic shock absorbers, anti-roll bar.
Tires	7.60 x 15 tube type whitewalls.
Wheels	Pressed steel discs, drop-center rims, lug-bolted to brake drums.

WEIGHTS & MEASURES

Wheelbase	124.0 in.
Overall length	208.0 in.
Overall height	60.4 in.
Overall width	77.6 in.
Front tread	58.5 in.
Rear tread	55.5 in.
Ground clearance	8.1 in.
Curb weight	3660 lb.

CAPACITIES

Crankcase	7 qt.
Cooling system	19.5 qt.
Fuel tank	20.0 gal.

FUEL CONSUMPTION

Best	14.0 mpg.
Average	10.3 mpg.

PERFORMANCE (from **Motor Trend**, 8/52 test of 145-bhp Hornet sedan):

0-30 mph	5.2 sec.
0-40 mph	7.8 sec.
0-50 mph	11.3 sec.
0-60 mph	16.8 sec.
0-70 mph	20.1 sec.
Standing ¼ mile	20.2 sec. and 70.2 mph.
Top speed (av.)	99.2 mph.

I blame the transmission's pre-overhaul roughness for causing the right rear axle to snap 5000 miles later. Ironically, the R&T Mercedes similarly needed a right rear axle replacement. It didn't snap, but some lock washers omitted at the factory let it slip out of its wheel. Mercedes again fixed R&T's axle under warranty, but in the Hornet's case, I had to pirate another axle from my single remaining parts car. I took out the entire axle assembly from the parts cars, removed the axle shafts and third member, and used these in the Hollywood. The rest of the parts-car differential became a potters wheel for the kids.

You might wonder how the Hornet axle actually snapped. I was driving about 25 mph along a 2-lane city street and wanted to pass another car. When I kicked down from fourth to third, the engine revved freely. There was absolutely no connection between the engine and the rear wheels. I thought at first a spline had let go. I've heard since that snapped axles in Hornets aren't too uncommon and that for racing, Hudson used shafts as thick as your wrist.

The third major breakdown in 12,000 miles took place on a trip back from L.A. last fall. Driving north on U.S. 99 at about 75 mph, I suddenly felt the car slow down. I thought at first that I'd hit a strong headwind. I didn't think much about it and simply pushed harder on the gas. Next thing I noticed was the temperature gauge creeping up toward H, so I backed off to 55 mph. The temp needle held steady at that speed, but whenever I'd try to go faster, the engine would start to heat.

I drove along that way for about 60 miles, when I had to stop for gas. Idling at the first light off the freeway, I felt the car shake as if the engine were missing on one cylinder. It was. I hoped it was a bad plug, but a quick check at the filling station showed me it wasn't. I nursed the car home at 50 mph (this was *long* before the gasoline crunch). At home, a compression check confirmed what had happened. That first "headwind" was a slowing down due to one valve no longer sealing. Why did it burn? Probably because after Jim and I overhauled the engine, we adjusted the valves cold but never reset them hot. Lesson learned, and that surely wasn't any fault of the Hudson's.

Jim Holmes performed the valve job. He replaced not one but three bad valves. The valve job cost me very little ($24.75), but with waiting for valves, gaskets, and then with some bad luck toward the buttoning-up process, the Hornet was out of commission for no less than six weeks. That's a hazard of owning an obsolete make. Luckily I had a backup car—my Camaro—so I wasn't in a hurry to have the Hudson back.

You understand, of course, that this repair and most of the others weren't performed in commercial shops. If they had been, the cost would have been greater but the downtime might have been less (although not necessarily). The Hornet's other, less major repairs are listed in the chart on the facing page.

Driving Around Town

People tend to smile when they see the Hornet. It gives them a little lift, I suppose, which in turn gives me a little lift. Many are curious to inspect it, and one of the car's hazards is its stigma as an attractive nuisance.

Other drivers crane to see it. I've had quite a few people stop me as I'm driving down the street.

One fellow in a red Chevy pickup cut a U-turn in the middle of Miner Av. to catch up with the Hudson. He motioned me to stop. I did, and I knew what he was going to ask. "Want to sell it? Is it for sale?" I said no, but we stood there and talked for 15 minutes about Hudsons.

Lots of people want to buy the car. They'll zoom up behind me on the freeway, heave alongside, and I can read the driver's lips. "Wanna sell it?" I shake my head no, we wave, and off he goes again.

Sometimes on the freeway, people will want to play games. Kids, especially. They'll pass and then sit in front of me, watching their mirrors. I'll pass them once, and if they pass me again after that, I slow way down. Eventually they get bored and go away.

On the other hand, it's a fantastic filling station car. Attendants leap to it. It's hard to beat for making friends fast. If ever I need a favor at a gas station—directions, a roadmap, to use the phone, john, or lube rack—no problem. It's a great conversation starter with just about everybody. People loosen up. I've watched little old ladies walk all the way around the car. People tell me about the time they or their uncles or their neighbors owned Hudsons. "Now there was a fantastic car."

Superwide armrest separates kids, can serve as writing table. Check thin pillars, ample head room.

How Is It On Long Trips?

When SIA began this experiment back in mid-1972, I figured we (the staff) would put about 15,000 miles a year on the car. It hasn't worked out that way. I'm now the only staffer who drives it, and it took me 18 months to put on these 12,000 miles. Just over half that mileage came in short, in-town trips. The other half involved longer hauls: five runs back and forth to Los Angeles, each averaging 1000 miles; two to Reno (500 miles each); and an unrecorded number to San Francisco, Oakland, and the Bay Area. I use the car for pleasure as well as business, and occasionally my wife and I have taken friends in it to restaurants and social affairs. Mostly, though, it's a business car.

It's on the freeway at sustained high speeds of 70-80 mph that the Hornet feels really at home. All Stepdown Hudsons share a reputation as long-distance runners. This car's effortless long-range cruising and its great comfort are the best things about it. I can spend 5½ hours behind the wheel on a hot summer day and still step out refreshed, even without air conditioning. Seating position, wheel height, good ventilation, and the gentle but firm ride all contribute.

Going to Reno, U.S. 80 over the Sierra crosses Donner Summit at 7227 feet. I've driven this route many times in other cars, and all but the Hornet have pooped out near the top. Even R&T's Mercedes elicited this remark, "At 8000-ft elevation the car had lost all its zip. . . ." But the Hornet hadn't. I don't believe the Hudson's speed has ever fallen below 65 mph on U.S. 80 over the Sierra. Same over the Grapevine going to L.A.

Hornet poses at Donner Pass on way to Harrah's, Reno. High elevations don't leave it puffing.

Handling is superb. There's neither over- nor understeer, and except for some wheel fight, fast cornering presents no problem. I've never broken the tires loose despite some very quick maneuvers in which I tried to do just that. Highway bends marked for 50 mph can be taken at 75 in comfort, with very little sway or roll. In some cars, too, fast cornering makes for driver confusion—too many different, unpredictable things coming up too quickly. Not with the Hornet. Fast turns are as easy as long straight stretches.

I do want to mention the tires. You might remember that when I first put the Hornet on the road, it had a set of 12-year-old, new-old-stock U.S. Royals—2½-in. whitewalls. I didn't trust these, so I put on four new Lesters. The Lesters haven't given me a bit of trouble. They've been over the freeways in 110° heat, have never had a puncture much less a blowout, I haven't babied them a bit, and they still show absolutely no wear. Tread depth looks unaffected by 12,000 miles of pretty hard driving.

What I do miss in this car is power steering. I thought perhaps I'd install it at one time, and I looked for a Hudson parts car that had it. Never found one, probably because at $177 new, power steering wasn't a popular option. The Hornet's unassisted steering is heavy and makes tight parking a real tussle. Most women couldn't fit this car into a short spot. Most men can't. I'd still like to install power.

What's it Like Inside?

For a car on a 124-inch wheelbase, there's precious little leg room in the back seat. Plenty up front but not much in the rear. My children find the seatbacks almost too heavy to push forward, and the seats' weight tends to catapult passengers rearward when they get in. Everyone trips at first on the perimeter frame (step up and then down), even though many newer cars are also "stepdown" by now. The huge doors tend to be awkward, especially in tight parking slots. The doors don't always shut without hard slamming, but luckily they rattle if they're not closed all the way.

What I like about the interior are the many little touches—things you don't find in cars nowadays. There are *five* domelights that come on with the doors, so the interior is like surgery. The seatbacks have big map pockets plus thick robe cords. There's a wide center armrest in the rear seat—the kids love it.

Both benches are fantastically wide. People marvel at how huge the front compartment is. Hudsons were among the first cars to have front seats wider than the car was tall. The seats are also very deep and buoyant.

Most Stepdown Hudsons had windows like tank slits, with fair-sized blind spots at the rear quarters of sedans and regular coupes. Luckily, the Hollywood hardtop has vast expanses of glass and almost no pillars to cut vision. By today's standards, the roof stands awfully tall. There's half a foot of air between the top of my head and the roof.

Two things do bother me about the interior. One is glare from those great expanses of chrome on the instrument panel. It looks nice, but it's dangerous when driving east around 5:00 p.m. The other bother is that the side windows don't seal well at the top. They whistle at freeway speeds.

Great care went into building Hudsons. I've searched for glue runs, sloppy mastic, shoddy fit and finish, and I haven't found any yet. Having replaced windows and even fenders, I find that workmanship is as faultless in the Hornet's hidden areas as on the surface. I don't say that Hudson was alone in its standard of quality. I believe all American cars were more carefully put together in those days.

Depreciation & General Expense

When SIA announced back in June 1972 that we'd try this experiment, we valued the Hornet at $850, despite having $1121.10 in it at that time. I now value the car at $1250, an appreciation of $400, and I feel I could easily get that for it today. (I'm writing this on Dec. 16, 1973 and I realize that by the time you read these words, the nation's economy and old-car prices might dictate an entirely different value.)

Editor drives this car mostly for business but also uses it for family and social functions.

Freeway cruising used to be an easy 70 mph, but no more. Chromy dash panel sometimes dazzles.

By totaling initial price, repairs, replacements, and refurbishments only, I find I have $1695.45 in the car. That means that despite appreciating $400 in one year, actual out-of-pocket loss was $445.45. We can call that depreciation, although technically it's normal expense and investment. If I don't put more into the Hornet, I can probably recoup the investment later on.

R&T's Mercedes depreciated $3377 in 24,000 miles of driving. Depreciation was, in fact, the M-B's largest single expense. And it's the largest expense of owning any new or newer car today.

By way of summary, I feel that the Hornet has much more than a low cost-per-mile figure going for it. In the first place, the owner of a brand-new car can also expect a couple of weeks of downtime. My POPULAR MECHANICS Owners Reports confirm this. And if new-car owners had to pay for repairs (warranty doesn't cover everything), they'd likely spend as much as I did in 12,000 miles.

As for gas mileage, most full-sized 1974 cars don't do much better than 10.3 mpg. The Hornet, though, ought to.

By eliminating the Hornet's two major drains (poor mileage and high repair expenses), I say that a 20-year-old car in good condition makes an incomparably better investment than a new one. It does even *with* the repairs and low gas mileage. Besides, consider the intangibles—the pleasure and enjoyment of driving a car like the Hornet which, despite its age, will still be breathing fire long after this crop of cars is gone and forgotten.

Our thanks to ROAD & TRACK *Magazine, Box 1757, Newport Beach, Calif. 92660; members of the Hudson-Essex-Terraplane Club, Peter Booz, Treasurer, 23104 Dolorosa, Woodland Hills, Calif. 91364; Jim Holmes, Stockton, Calif.;* POPULAR MECHANICS, *224 W. 57th St. New York, N.Y. 10019; and Jack Clifford, Clifford Research & Development Corp., Costa Mesa, Calif.*

At the 1953 Motor Show at Earls Court several American car manufacturers were represented. Among them were Nash, Kaiser, Willys-Overland, Pontiac, Buick, Cadillac, Oldsmobile and Hudson.

Hudson, whose UK headquarters at the time were on the Great West Road, Chiswick, London W4, had three examples of their six cylinder Hornet Saloons on show, according to the official catalogue.

These 17ft 4in monsters must have represented an incredible sight. Looking like factory 'lead sleds' with their low roof lines, 6ft 5½in overall widths and covered rear wheels.

One of those three cars is still at large and at the time of writing was up for sale. It belongs to Bette Hall from near Mildenhall in Suffolk.

Her car 'XMM 2' was bought by her late husband Wally in 1989. Fate was to declare that his enjoyment of the car, his first American, was to be shortlived. He died early in 1990 after only six months with the maroon and white leviathan.

Bette now wants to sell the car. It's not that she's scared of driving it. Though petite she holds a heavy goods vehicle driving licence, so the Hudson is small potatoes in that

SHOWBOAT

respect. No, it's because her life has changed so dramatically that she is moving house to make a fresh start.

Because of her husband's death she has also stopped driving her two lorries. Instead she acts as manager, employing two drivers.

Her work involves collecting and delivering American furniture for American air force bases in this country. US airmen of certain ranks are allowed to have their own furniture shipped across so that they feel more at home while serving in the UK.

When Bette and Wally went to buy the Hudson they were amazed to discover it was the very example featured in the book 'Street Dreams: American Car Culture from the '50s to the '80s' by David Barry. There on pages 24-25 the car is pictured in full colour.

"I believe the colour is original," said Bette. "We bought it from someone in Bristol. He owned it for five or six years and said he used it all the time, taking it to shows etc."

The car was bought off the show stand in 1953 by a doctor from Brighton. One went to Finland, and what happened to the other car is a mystery.

There's a plaque on the ashtray to say it had done its first 100,000 miles on such a date on such a road (March 3 1959, 1½ miles outside Torquay on the Newton Abbot road).

"It was later bought by the president of the Hudson Owners Club, and the Bristol man bought it off him."

The Hudson was very busy in the

Exhibited at Earls Court thirty eight years ago, this Hudson show car survives intact. Bob Crookes reports

first ten years of its life. In the last 30 it has only covered a further 35,328 miles according to the odometer.

In running order it has been left in Bette's garage since the MoT ran out in June '90. Once the six volt battery was charged up the six cylinder engine fired straight away and the three speed manual gearbox with column change was engaged and the Hornet slowly eased out into the sunlight.

"I can't get the seat far enough forward," said cheroot smoking Bette, adding: "you can't park it in Tesco's car park. It doesn't have enough lock on it!"

The Hudson was built for wide expansive roads and wide expansive times in a country where due space was accorded the kings of the road.

Inside this six seater you find green cloth covered bench seats front and rear, a deep chromium trimmed facia which reaches forward to the split screen, the two halves of which are angled to meet at the centre point.

The impression given to the driver is like being at the helm of a classy motor cruiser. Many American cars have rather cold interiors. The Hornet is entirely different. It is warm and friendly. The peak and low screens combine to give occupants a degree of privacy which adds to the glow.

The whole car is like a factory alternative to a roof chopped Mercury lead sled, with extra windows at the back. These are hinged so that they can be opened.

Being an export model, this Hudson features right hand drive. Presumably the rest of its specification is the same as the market models.

According to the official 1953 show catalogue the car's price then was £2025 plus £844 17s 6d tax (total £2869 17s 6d) — expensive for the time when you consider that the basic price for a Jaguar MkVII saloon was £1140, a Humber Pullman Limousine £1395 and Ford Zephyr Zodiac £600.

On the other hand, several of the other American cars on show were dearer, the most expensive listed being £2900 basic for Cadillac's Special four-door saloon.

The three Hudson Series 7D four door saloons were reported by the catalogue to be on stand 138, and part of the catalogue's description of the car includes the following:

Engine: 96.8mm bore x 114.3mm stroke, 5048cc developing 160bhp. Compression ratio 7.5:1, miracle dome aluminium head, side valves, dual downdraught carburettor fitted with air cleaner and automatic choke, coil ignition, six volt 100amp-hour battery. Clutch: single plate cork inserts running in oil. Needle bearing mounted worm and roller-type steering.

Three forward speeds. Transmission ratios: standard first 2.88:1, second 1.82:1, third 1.1:1, reverse 3.50:1. Semi-floating rear axle with alloy hypoid gears.

Suspension: front independent coil, rear semi-elliptic: drop-centre type wheels with 7.60x15 white sidewall super cushion tyres.

High quality ribbed Bedford cord cloth upholstery, air foam cushions, functional air scoop bonnet ornament: bench type front seat, and rear seat with 16in centre armrest: deluxe robe hanger, pocket on back of front of seat, coat hooks, large parcel compartment with lock, side armrest to front and rear seats, assist straps in rear compartment.

Equipment includes: overdrive, radio, weather control heater with remote control, automatic ejection cigarette lighter, window and wing vent shades, direction indicator electric clock, petrol gauge, teleflash warning signals, twin air horns. Fuel tank capacity 16½ gal, petrol consumption 15-17mpg, max speed 98-103mph.

According to The Autocar report of the 1953 Motor Show there were four Hudsons on stand 138. The report said:

Four examples of the Hornet are displayed this year. All cars have identical specifications with the exception of the colour schemes. . . Other modifications include the use of a one-piece windscreen.

All the models are finished in attractive two-tone colours: light grey with a red roof, two shades of blue – both with blue Bedford cord upholstery – while the other two cars are in Roman bronze and black, and red and black.

Presumably the car we are dealing with on these pages is the red and black one. The reference to one-piece windscreen conflicts with the car pictured on these pages.

The Hornet was introduced in 1951 with four-body styles. It was based on the 'step-down' chassis design introduced by Hudson in 1948. The unit body was titled thus because the floorpan was dropped and surrounded by frame girders. As well as making it strong and safe from a crash point of view, it radically lowered the centre of gravity and enhanced handling.

The '48 car came with an inline eight or new 262cu in Super Six power unit giving 121bhp at 4000rpm, only 7bhp less than the eight.

The unit was enlarged to 308cu in for the 1951 Hornet and was said to be the largest L-head six ever built. In stock form it gave 145bhp at 3800rpm and responded well to tuning. Famous Hornet race driver of the era Marshal Teague claimed he could get 112mph from a Hornet certified as stock by the AAA or NASCAR.

By using special parts including Twin H-Power first seen in 1953 — twin carbs and dual manifold, claimed to be the first such manifold on a six — the Hornet could be boosted to over 200bhp.

From 1951-54 the Hornet was all but unbeatable in AAA and NASCAR events. In 1952 Teague won 12 of the 13 championship stock car events.

The step-down series had been a spectacular success for Hudson, but the company had invested heavily in an ill fated new model, and the history books tell us Hudson could not afford to update existing models with new body styles.

The names Hudson and Hornet disappeared in 1957 when American Motors also decided to drop the Nash name in favour of Rambler.

The Hudson factory may be a thing of the past, but some of its cars live on and this one is for sale. It can surely be one of only a select number of four door versions still running in this country. Owner Bette Hall is interested in hearing your offers. Call her on 0842 861811.

Above: Mighty straight six flathead is fired by six volt battery and coil ignition

Previous owner Steve Passmore, of Old Sodbury, Bristol, who had the car for seven years confirms that this is indeed a 1953 Motor Show car.

He made extensive enquiries in America about the model to find as much about it as possible after buying it off the Hudson historian in this country.

Steve says the car was bought at the show by a Harley Street doctor. He says there were six Hudsons on display — three 1953 models and three 1954 models.

The specification we carry on these pages, taken from the official 1953 Motor Show catalogue, must apply to a '54 Hornet, he argues, because it refers to Bedford cord interior, whereas the model shown on these pages has what he terms 'green brocade' seat covering, available only during 1953 as an alternative to leather.

Though there are plenty of '52 Hornets in this country, says Steve, he knows of only one other 1953 model. "They were never imported into this country in 1953," he stated.

The seat material used in this show car is now impossible to get hold of, cautions the former owner who bought the original front window peak from America.

He reckons the original owner was responsible for several tasty modifications done very professionally. These include reinforcing the front bumper to take the weight of spotlights, fitting spotlight mounting brackets that look as though they came from the factory, and an electrically heated rear screen.

He sold the Hornet to concentrate on his 1930s American cars. He still has some spares including an engine which he feels ought to remain with the car. Wally Hall never got round to collecting them, so Steve would like whoever buys the car to get in touch. Call him on 0454 313467.